F.J. Milford

Mine Warfare Technology

Louis C. Gerken

Mine Warfare Technology

ISBN Number 0-9617163-1-4

Publisher: American Scientific Corp.
3250 Holly Way
Chula Vista, CA 92010

Author: Louis Gerken

Library of Congress Cataloging in Publication Data
Main entry under title:

Gerken, Louis C. 1925-
Mine Warfare Technology

1. Mines, Technology, Historical, Aircraft, Ships, Submarines, Tactics, Operations, Sensors, Explosives.

ISBN 0-9617163-1-4

Printed in the United States of America

Prologue

This book will take a broad look at the historical and technological aspects of mine and mine-countermeasure warfare and equipment of the past, present, and trends of the future.

Included are some human aspects and anecdotes that should make this book more interesting to the reader.

The Author

Table of Contents

Chapter 1	Introduction Historical Evolution of Mines & Mine Warfare	1
Chapter 2	Minelaying, Minesweeping and Mine Countermeasure Platforms Submarines Surface Ships Aircraft	21
Chapter 3	Typical Mines of Various Nations	81
Chapter 4	Minehunting Sonars Mine-Countermeasure Sonars	135
Chapter 5	Remote Mine Neutralization Systems	161
Chapter 6	Mine Warfare Navigation/Plotting Equipment	195
Chapter 7	Degaussing of Vessels Vessel Construction Techniques	207
Chapter 8	Typical Mine Warfare Operations	217
Chapter 9	Outlook for the Future	245

Dedication

To those officers and men of all navies
who have fought the mine and mine-countermeasure
battles of past wars and those who are actively
engaged in present efforts with "Weapons that Wait."

CHAPTER 1

Historical—Evolution of the Naval Mine: Discovery and Development

MINES BEFORE WORLD WAR I

The first appearance of mines in the history of naval warfare reportedly was in 1585 when Gianibelli, an Italian engineer employed by the Dutch, blew up a Spanish floating bridge that was blockading Antwerp. At that time Antwerp belonged to The Netherlands. The Spanish commander Parma was thrown into confusion as he lost his bridge and about a thousand men. Gianibelli's plan was to experiment with two small vessels packed with gunpowder and equipped with a clock mechanism which operated a flintlock firing device some time after being set. He floated two ships down to the bridge at night, masked by 32 fireships sent out as a diversionary attack to deceive the Spaniards into thinking his mine-vessels were duds. One was, but the other one worked.

In 1588 fear of mines caused a break in formation in the ships of the Spanish Armada leaving from Calais, France toward its eventual defeat by the British. The British admirals sent fireships down into the Spanish Armada exiting Calais. The Spaniards recalled the "Fireships of Antwerp" and in their haste to run away they lost several ships by collision and were panicked by the dummy "mines."

In 1775 David Bushnell, an American and a graduate of Yale, experimented with a weapon called a torpedo. The Torpedo fish was found off European coasts. (The root Latin verb, *torpere*, means to stun.) This name was used through most of the 19th century for the weapon we now call the mine. The advent of the self-propelled torpedo caused the name of the stationary device to be called a mine.

Bushnell, who invented America's first submarine, invented the sea mine in 1776. (See Figure 1.1) He first floated wooden kegs filled with gunpowder and fitted with contact firing devices down on the British in Philadelphia harbor during the Revolutionary War. Though not one struck a British ship, the crews panicked. They fired into the water, fighting what was called the "Battle of the Kegs."

In 1777, at the request of George Washington, a number of these mines in kegs were set adrift by David Bushnell. In an attempt to make contact with the ship's hull more certain, Bushnell invented the submarine. Figure 1.1 shows that it was driven with hand power by cranking a screw. It carried a payload of a "mine" (sometimes called a "torpedo.") An auger was used which could be screwed into the ship's wooden hull to

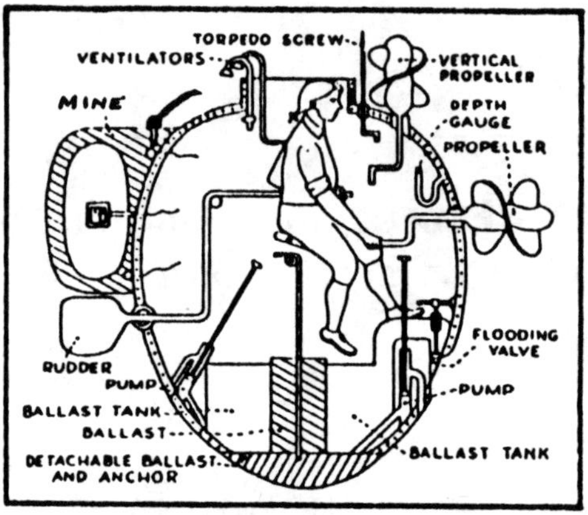

Figure 1.1. At top, David Bushnell's submarine carrying a mine aft. Below, model of Bushnell's "Turtle" made by Major M. Gray in the British Royal Navy Submarine Museum, Gosport, Great Britain.

be attacked. This vessel was called the Turtle because the oak hull was similar to two tortoise shells mated together. It was 7.5 feet long and 6 feet deep, and bound with iron bands. An attack is alleged to have been made on the British ship *HMS Eagle* by Sgt. Ezra Lee in New York Harbor during the American Revolution. The attempt was unsuccessful since the ship's hull was coated with heavy copper sheathing and the Turtle's auger could not penetrate it. The auger screw at the top was attached to the mine powder charge delayed-action explosive by a lanyard. So, the first limpet mine was born.

Robert Fulton, of steamboat fame, designed several sea mines between 1797 and 1812. He tried to sell them to France, Great Britain and then to the United States, but was unsuccessful. Many of his mines were successful when tested but he received little encouragement from customers. His mine was essentially a spar-torpedo, an explosive put on the end of a long pole and rammed at the hull of an enemy vessel.

In 1812, Samuel Sommering experimented with telegraph apparatus using wire insulated with India rubber and varnish. He succeeded in telegraphing through some 10,000 feet of cable. Baron Pavel Schilling von Cannstadt, a member of the Russian diplomatic mission in Munich, visited Sommering and became aware of his experiments. Since war with France was imminent, Schilling carried out a demonstration of detonating powder charges across the Neva River in St. Petersburg (now Leningrad) in Russia. After Baron Schilling's death in 1837, his efforts in galvanic mine development were reveiwed by the Russian Imperial Academy of Sciences. Czar Nicholas I appointed a young Prussian emigre, Moritz von Jacobi, as head of a working group to carry out galvanic mine developments for some fifteen years before the Crimean War (1854-56). The Russians actually did deploy some mines during that war at Kronstadt to defend St. Petersburg, although the hostilities actually took place a thousand miles away on the Crimean peninsula. Jacobi is best known for the invention and development of a device for firing gunpowder. This was a chemical method of generating heat and consisted of a glass tube filled with sulfuric acid and imbedded in a mixture of sugar and potassium chlorate powder.

In 1842 Samuel Colt of Connecticut devised experiments with mine hardware at the age of 15. He also invented the revolver pistol. Colt adapted an electrical firing device to his mine designs, developed and built a successful sheathed underwater cable and conducted numerous spectacular and successful demonstrations of his system. Some of these were against moving craft at considerable distances from the firing point. The U.S. Navy was not sold on the idea or practicality of his mines.

In 1846, at the age of 27, Werner von Siemens was appointed to a commission establishing an underground telegraph system in Prussia. During the Schleswig-Holstein War in 1848 (with Denmark), von Siemens collaborated with his brother-in-law, Professor Karl Himly of the University of Kiel in designing and laying a field of galvanically controlled mines in the approaches of Kiel, that effectively discouraged Danish naval bombardment of the Kiel seaport. The mines were wine cask mines, so it can be presumed that this was the first controlled moored minefield in history.

In 1854-56 the Russians planted mines during the Crimean War at Sevastapol, Sveaborg and Kronstadt. These used the Jacobi fuse—a kind of chemical horn firing device.

In 1857-1858 the Chinese used mines in the war with the British.

During the French-Austrian War in 1859, electrically controlled mines were used in the harbor of Venice. The position of the mines relative to ships was determined by means of a camera obscura in which a chart marked with the mine plants could be compared with the reflected images of ships on the surface. Perhaps this was an early indication that navigation accuracy was needed to keep track of mine locations in the fields. The mines contained up to 450 lbs. of guncotton rigged to be fired electrically.

The Hertz horn, which was to prove so valuable in firing contact mines, evolved right after the German-Austrian War of 1866. The invention consisted of an electrolyte (potassium bichromate solution) in a glass tube sheathed in a soft metal horn, usually lead, which stuck outside of the mine. When bent by contact with a ship, the glass would break, and the electrolyte would complete the circuit in a battery which would fire the electric detonator. This became standard for contact mines.

Up to the middle of the 19th century, sea mines were used in several wars in Europe and Asia, but with little effect. The weaker of the opponents was often forced to the greatest ingenuity. At this stage of development, the weapon was both a mine and a torpedo. There was a change in tactics taking place that made this possible, from fighting land engagements on board ships (using the techniques of grappling, boarding parties and ship capture) to fighting the ships where the goal was destruction of the vessel itself. Gunfire was the method of accomplishing this and the greater vulnerability of a vessel below its waterline was not exploited because underwater weapons had not reached a practical state.

During the American Civil War mines were used on a relatively large scale. The Confederate Navy was the underdog and had to use their ingenuity to correct this imbalance. They needed a weapon that could be quickly and cheaply produced to protect their harbors. Several types of mines were developed and put to use. (They were still called "torpedoes" at that time.) These defensive mines, demijohns filled with gunpowder, were ignited by a friction primer. The cases were laid in pairs connected by a wire passing to friction tubes. When a ship struck them, it pulled the wire and exploded the mines. The Confederates also developed a contact mine which did not have to be carried against a ship but waited and exploded when the ship came upon it. The "Singer" mine contained an air chamber for buoyancy and a friction firing device which was set off when the mine was jarred. This Singer was the brother of the sewing-machine manufacturer. His mine was credited with being the most successful Confederate mine device. It was a hand-planted, moored mine with a charge of about sixty pounds of black powder. It carried a heavy cap in its upper surface that would be knocked off by a ship running into it. When it came off, it released a spring-driven plunger that struck a fulminating charge—thus exploding the mine. Another type of mine was horned, with as many as five fuzes jutting out at various angles. When broken, the horns spilled sulphuric acid on chlorate of potash mixed with powdered sugar. This generated enough heat to set off the charge.

Confederate submarine builders were Captains J. McClintock, H. Hunley, and a civilian engineer, Mr. B. Watson. They completed the submarine *Pioneer* in 1862 at New Orleans, but several men drowned during initial operations. Before it could be sent into action against the Union fleet under Admiral Farragut, New Orleans was captured. The three builders escaped and made their way to Mobile. In 1863 Capt. Hunley, using his personal, funds constructed a new submarine. With the help of Lt. J. Alexander, an engineer officer, Hunley built a submarine 30 feet long and 4 feet wide. A three-bladed propeller was driven by a crankshaft turned by eight men. (See Figure 1.2.) An old boiler was used as a hull. To give maximum concealment when running on the surface, the craft was designed with only a few inches of freeboard. Since there was no periscope, the submarine was navigated by the captain standing in the forward conning tower-hatch with his head out. After successful trials in Mobile Bay, the vessel was shipped by rail to Charleston, South Carolina where the *CSS H. L. Hunley* was prepared for action. While attempting to engage in enemy action, the Hunley sank twice with the loss of life, including Hunley himself.

In 1864 Lt. Alexander made another attempt against the Union ship *USS Housatonic*

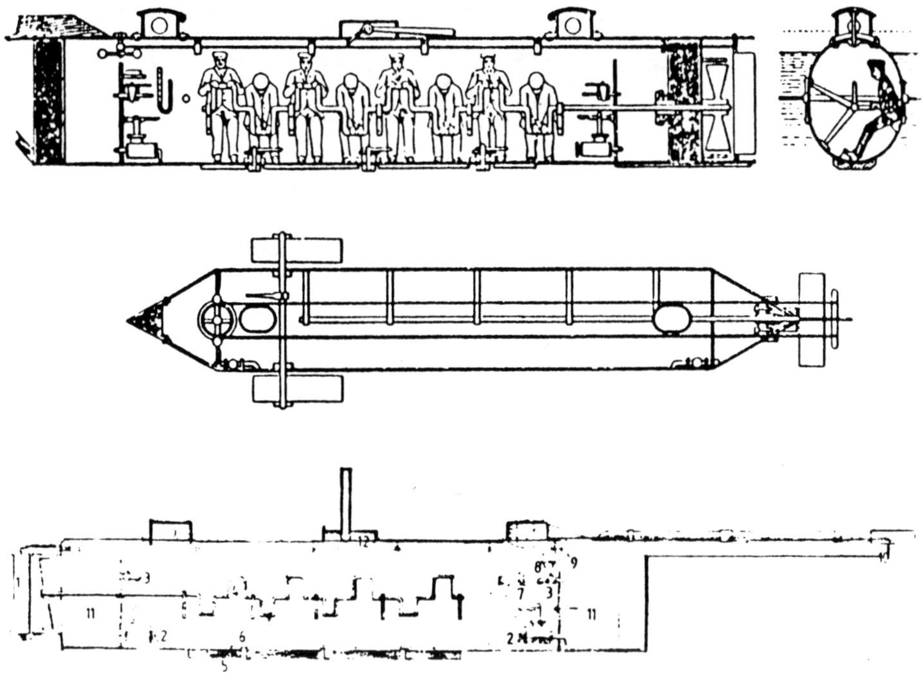

The Confederate *Hunley* 1863

1 rudder
2 pumps
3 sea cocks
4 propeller crank
5 cast iron keel
6 release bolts
7 mercury guage
8 compass
9 steering wheel
10 spar torpedo
11 ballast tanks (open)
12 air box

Figure 1.2. ***CSS Hunley*** **Submarine**

near Fort Sumter, South Carolina. The Confederate *Hunley* sank both the *Housatonic* and herself with the one spar torpedo (or mine) (See Figure 1.3). The *Hunley* was under the command of Lt. George Dixon of the 21st Regiment of the Alabama Light Infantry, gained the distinction of sinking the first ship by submarine action in wartime. If the hatch on the submarine had been shut and if some means had been devised of detonating the mine on the spar from inside the hull, Lt. Dixon and his men might have been saved. Lt. Dixon rammed the *Hunley's* spar torpedo into the side of the steam

Figure 1.3. History's first successful submarine attack was made during the Civil War when the Confederates destroyed the *USS Housatonic* with a spar mine.

sloop of war *USS Housatonic*. The sloop sank with no loss of life. The *Hunley* also sank with the loss of her crew of five.

At Mobile Bay, Alabama in 1864 a field of 80 mines was laid by the Confederates. The mines were laid in a triple line, staggered so that if a ship passed between mines in one row it would meet a mine in the very next row. The Union Naval Commander Admiral Farragut ignored the threat of the mines and gave the command for a squadron of his ships to attack Mobile. The *Tecumseh* led the way. Before she was near enough to shore to use her guns, she struck a mine. After a great explosion, the hull tilted forward and she quickly sank. Officers and most of the crew were lost. Another ship saw more mines in the water, and began to signal and also started to alter her course. But Admiral Farragut ordered, "Damn the mines, full speed ahead." Mines bumped against other ships but none exploded. Many of the Confederate mines had corroded and were rendered useless. Others perhaps were affected by the wave action and did not function properly. When the mine field was finally cleared, there were a number of casualties.

During the American Civil War, the Union losses were quite high. Thirty-two Union ships were mined, of which 27 were sunk. Only 9 Union ships were sunk by Confederate gunfire.

In 1882 hundreds of steamers were tied up outside the Suez Canal because of the reported planting of mines during the Arab-Pasha uprising. An Italian Navy Capt. Morin (later V/Admiral Morin) was in command of an Italian warship. Rather than wait, he stated that he did not think that the Egyptians had the skill to lay mines properly, and if they did, the mines had been laid so long that they were probably ineffective. Capt. Morin may have heard of Admiral Farragut's bout with mines and he likewise decided to steam ahead. Fortunately, he encountered no mines, or at least no damage, so the other steamers followed suit.

The Russians used sea mines during the latter half of the last century in the Crimean War. In 1887 a Russian mine sank a Turkish monitor ship in the Danube river.

In 1898, during the Spanish-American war, the U.S. battleship *Maine* sank in the Havana, Cuba harbor following an explosion attributed to a Spanish mine. However, it was later determined that the explosion that sank the *Maine* was caused by an internal explosion and not by a mine. A fire in a coal bunker set off a powder magazine in an adjacent compartment that was not well insulated in the ship. The Spanish did employ mines off Cuba and likewise placed them at Manila and Subic Bay in the Philippines. These mines were either avoided or were ineffective, at they produced no casualties. But the believable allegation that the Spanish had somehow subversively blown up a U.S. battleship moored peacefully in Havana harbor was enough to incite the U.S. public and Government into action and war was declared.

The next important use of mines occurred during the Russo-Japanese War of 1904-05. Mines played a decisive role in this war. The success of mines in the U.S. Civil War had been observed by the European powers and Japan which resulted in widespread development of the mine. It is not surprising that the Russians would rely on defensive mining outside their Pacific ports, as they had been developing mine concepts and uses since the days of Jacobi and before. However, it is surprising that the Japanese emerged at that time as a major naval power. They had very modern mines and used them offensively. The Japanese had learned to cast-load the mines with picric acid and were credited with having a new secret super-explosive. The Japanese were able to lure a Russian force over their minefield with great losses to the Russians.

The Russians used mines primarily for defensive purposes. They laid a field of about three hundred to protect the Port Arthur area after a successful Japanese attack. These mines prevented Japanese ships from coming in to give fire support to their troops on the land. In this war, counting both sides, three battleships, five cruisers, four destroyers, two torpedo boats, one minelayer, and one gunboat were sunk by mines while other ships were severely damaged. The Japanese minelayer *Yenisei* of 2,500 tons blundered into her own minefield and was blown up by a mine that she had just laid. This demonstrated the need for good navigation, or at least a delayed arming mechanism so that the minelayers could get clear. The Russians lost their 11,000-ton battleship *Petropavlovsk* which hit two mines and went down while crossing a known Japanese minefield. The ship struck the mid-point of a cable connecting two Japanese mines causing them to be dragged simultaneously into her port and starboard sides. This incident initiated an urgency in the development of a countermeasure to mines that remains undiminished to the present day. The Russian Admiral Stepan Osipovich Makaroff lost his life in that event. He refused to consider the mines dangerous and would not change course. The Japanese attacked, but sometimes as a ruse for laying their own mines. They lured three Russian battleships and four cruisers out of Port Arthur through a Japanese minefield. The Russians steamed right through without incident but on sighting Admiral Togo's main squadron coming up over the horizon, the Russians turned back toward Port Arthur. However, on the return trip the Russians were not so lucky and they had grave losses. After the Japanese approached Port Arthur by land, the Russians withdrew by sea. They had to go through their own minefield, which they were not able to sweep beforehand, and had additional losses.

In order to disrupt Japanese shipping, the Russians laid mines in the open sea. However, this brought many protests from Western countries. Sometimes it appeared that a Japanese ship had struck two mines at once. This gave rise to a rumor that the Russians had a new and more powerful mine. Actually they were using the usual type of electromechanical mine that sometimes was able to explode a ship's own store of ammunition.

The Russo-Japanese War showed that mines were formidable weapons. The Russians sank more Japanese ships by mines than by any other form of attack. Japan lost two battleships, four cruisers, two destroyers, one torpedo boat, and one minelayer (the *Yenisei*, which struck her own mines). The Russians lost one battleship, one cruiser, two destroyers, one torpedo boat and one gunboat. This was a rather impressive record for the lowly mine.

After the Russo-Japanese War the Hague Convention of 1907 was held in The Netherlands to discuss ways of controlling the use of mines. However, there was much difference of opinion as to when, where and how mines could be used. To begin with, the British were unable to persuade the Germans and Russians to give up mining altogether. The conferees finally agreed to a number of provisions that were ambiguous in fact and were unenforceable even if there had been no loopholes.

MINES IN WORLD WAR I

In 1914, the memory of Port Arthur was still fresh in Europe. The British suspected that German doctrine relied upon mines, torpedoes and submarines to counterbalance the British numerical superiority. Both Germany and Great Britain heeded the events of the Russo-Japanese War and both undertook strategic and tactical mining on a large scale. During World War I, the naval mine emerged as the Allies' primary defensive weapon against the submarine. Great Britain used minefields to establish war zones that required most North Atlantic shipping and ships headed for the Baltic Sea to enter British ports for clearance and routing, assuring British control of shipping in the area.

Perhaps the greatest impact of mines was psychological. Commander-in-Chief of the British Grand Fleet Admiral Sir John Jellicoe advised the Admiralty office in a letter of his intention not to be drawn into a tactical trap set up by the apparent retirement of the German fleet.

> "The situation is a difficult one. It is quite within the bounds of possibility that half of our battle fleet might be disabled by underwater attack before the guns opened fire at all, if a false step is made, and I feel that I must constantly bear in mind the great probability of such attack and be prepared tactically to prevent its success."

Shortly after this the British had a taste of defeat by mines in the Dardanelles campaign against Turkey. Mines played a key part in one of the most disastrous campaigns of World War I, the British attempt to capture the Dardanelles.

Turkish forces, aided by the Germans, had planted defensive mine fields and, perhaps most critical for their success, were covered with heavy shore artillery from commanding positions. They used anchored contact mines set at different depths, mines controlled from shore, and drifters, which were set in the strong currents of the March flood. Operating against the Turks, the British Fleet sent their sweepers out at night.

However, Turkish searchlights picked them up, shelled them heavily, and made sweeping almost impossible. This action, together with the artillery bombardment from the shore, took place over a period of days. In covering the destroyer and trawler sweepers, the cruiser *Amethyst* was severely damaged from shell fire. The British tried for a month to pass the Dardanelles, but were not successful. The minefields, which were first regarded as only a annoying obstacle to the main naval mission, finally proved insurmountable.

As the British were relieving the French squadron, the battleship *Bouvet*, coming out of the line, struck a Turkish mine that exploded her magazines. She split wide open, turned over and sank, still going ahead, the water closing ove, the twisted, shredded mass of steel about three minutes later. Then the *Irresistible* was struck, then the *Inflexible*, both British battleships. The *Inflexible* got off to Tenedos where it was found that she had flooded twenty compartments and had a hole in her bottom 30 by 26 feet. The *Irresistible* was holed by another mine as she drifted helpless, her engines stopped. The battleship *Ocean*, coming to her aid, received a hit in her steering gear, and while running out of control struck a mine. Both ships, now being subjected to heavy bombardment from the shore, sank during the night. The Allied forces finally withdrew. Winston Churchill said of this action in 1915:

> "The operation of sweeping the area from which the ships were to bombard was the indispensable preliminary to any naval attack upon the forts. This was not achieved because the sweepers were not adequate. This led directly to the losses in the attack, and then indirectly to the abandonment of the whole naval enterprise."

Failure in the Dardanelles led to the disastrous British Gallipoli campaign that Churchill helped organize. It was hoped that amphibious forces could land and assault the upper land where the Turkish artillery was located. But this likewise proved impossible for the British.

During World War I the Germans conducted extensive mine operations, using both surface raiders and submarines for laying mines. They defended their ports with mine fields, and they set groups of mines where they could most harm the British on the high seas. The mines were of an indeterminate number, scattered at random distances in order to make sweepinng harder. As the weaker naval power, the Germans used mines offensively and made it apparent that they considered mines an important weapon. The German raider *Wolf* laid mines successfully in areas off the northern tip of New Zealand and along the route of the Atlantic liners. Near Tory Island the British battleship *Audacious* struck a mine, and in just four hours she sank. A German mine near Marwick Head caused the disaster to the British cruiser *HMS Hampshire* in June 1916. During a severe storm she struck a mine and sank in 20 minutes. All but 14 of the 600 men on board were lost, including Lord Kitchener, the British Secretary of War. In 1917 the British liner *Laurentic* hit a mine and went down with many of her complement.

Between 1914 and 1917 the British lacked effective mines. Only about one in twenty were reliable in firing or stayed at the proper depth. The primary batteries ran down, the contacts rusted, and the explosive deteriorated with the dampness. In 1917 a new British mine, the Mark H was developed and manufactured in quantity and the situation improved.

One of Great Britain's mine laying operations was the Coast Barrage, two lines 40 miles long, that was laid 12 miles off the coast of Belgium. This was meant to deny the Germans the use of the English Channel to exit into the Atlantic. The Allies laid 172,000 mines in the English Channel and the North Sea. Between 1914 and 1918 enormous fields were laid to bottle up the German submarines. In antisubmarine barrages, like that in the North Sea, mines were anchored at such depths that surface ships could pass safely over but submerged submarines would run into them. The construction of the North Sea Barrage was so deep and the barrage so long that a large number of mines and minelayers were needed. Most of the mines were laid by surface ships. The waters adjacent to Norway were not mined and left a loophole for the German submarines. However, Allied vessels tried to keep close watch along the coast. Figure 1.4 shows U.S. sailors assembling a mine at Inverness (in northern Scotland) for the North Sea Barrage in September 1918.

The famous North Sea Mine Barrage was laid commencing in June 1918 in the North Sea. Most of these mines were the familiar MK-6 type, designed for mass production and rolled off U.S. factory assembly lines. The 70,117 moored antisubmarine mines were laid in a belt 15 to 35 miles wide and reached 230 miles from Scotland to Norway. The British Navy laid some of them, but 80% were laid by the "Yankee Mining Squadron." The North Sea Mine Barrage was not fully completed before the war ended. Therefore it is not known how successful it would have been if completed. Clearance of these mines after World War I took over a year. The Mine Barrage possibly sank at least six German submarines and damaged many more. Perhaps the most important impact that these mines had was that the submarines that risked the barrage and managed to reach the Atlantic wasted valuable time and much of their diesel fuel because of the required evasive tactics.

German submarines placed 58 mines in six U.S. ports in 1917. The armored cruiser *USS San Diego* (CA-6) and four merchantmen were sunk. The battleship *USS Minnesota* (BB-52) was also damaged. As a result, several of the U.S. ports were closed and extensive sweeping was required to open them.

British submarine minelayers were credited with sinking a hundred enemy ships between 1915 and 1918. They operated in the Heligoland Bight off the coast of Germany in the North Sea.

Figure 1.4. U.S. Naval sailors assembling mines for North Sea Barrage.

	Warships	Auxiliaries	Merchants
German losses	102 (includes 40 submarines)	42	–
British losses	44	225	–
Allied losses	–	–	586 (over 1,000,000 gross tons)

In 1914 the Russian Admiral Nil Lai von Essen planned the use of mines against the Germans in the Baltic Sea.

At the outbreak of World War I the Russians laid some 2,200 moored mines at the entrance of the Gulf of Finland to protect St. Petersburg from attack from the sea by the Germans. These fields were supplemented by additional mines off the Latvian coast and along the German iron ore route from Sweden to Germany. As a result of these minefields, the Germans lost four cruisers and seven torpedo boats. These Russian mine operations were very clearly conceived and professionally executed. Unfortunately for the Russians these mine operation successes were not able to tip the balance of such Russian military disasters when they were beaten by the German Army at Tannenburg or the effects of the Revolution of 1917 when Lenin took over the government.

The most significant development of the Germans in their North Sea mine operations was the use of submarines as minelayers. These vessels were dedicated to this purpose. They were capable of carrying and launching eighteen, and later thirty-four, Type IV mines. These were launched through six mine tubes which were inclined aft from top to bottom to prevent mines from hanging up when the submarine was moving. A soluble plug was used to delay the release of the mine from the anchor by ten to twenty minutes so that the mine-laying submarine could get clear. Later in the war the Germans developed a delayed rising or a variably delayed rising mine which took from one to four days so that mines would appear later in fields that the British had already swept.

In 1915 six British E-class submarines were converted into minelayers. These had vertical tubes in saddle tanks. A total of 32 mines could be carried in 16 tubes.

Also in 1915 the Russian "Krab" was the first submarine minelayer. It used a rather elaborate system for laying the mines. They were planted through two stern-facing tubes and conveyed rearward by a conveyer belt.

During the last phase of World War I, magnetic and acoustic mines were introduced. During this period mine countermeasures became an important consideration because of the many mines laid. The mine had at last become accepted as an important defensive and offensive weapon in naval war strategy. Unfortunately, after World War I, in the years of peace that followed, the mine was all but forgotten. Being a somewhat unspectacular weapon it was not considered worth any great development effort and received very little funding.

MINES IN WORLD WAR II

For several years the U.S. had received warning that it might be brought into another war in 1939–1940. However, there was a very low level of effort in mine-warfare technology even though the Navy had embarked on an extensive shipbuilding program. There was a very small staff section in Washington covering the mine warfare organization.

Mines took on a new role in World War II. They started to be used as an offensive weapon. The submarine was used more as a mine layer and the aircraft was introduced as a mine layer. Influence-fired mines were designed. From small beginnings the mine warfare effort became larger in scope in keeping with all other wartime undertakings.

During World War II thousands of mines were laid for offensive purposes. Mines were designed to defend themselves against attacks of nature such as wave action and sea growth, since many were dropped in waters which we did not control. Mines also were designed to defend themself against sweeps by the enemy.

During the War, the British laid eleven mines in the Kiel Canal in North Germany. These mines were estimated to have resulted in the virtual loss of over one million metric tons of cargo to the German war effort.

Germany mined North American ports, planting about 350 mines from Trinidad to Halifax. Since it was considered more fruitful for the submarines to be at sea sinking ships with torpedoes, the Germans could not spare many of their U-boats for mining purposes. With the few mines placed, several important ports were closed for up to two weeks and required extensive sweeping and deployment of defensive fields.

Amphibious operations were used extensively during World War II. This necessitated assault sweeping at every beachhead, and the setting of defensive fields. This placed the mine forces at the front of these assaults.

A classic use of sea mines was "Operation Starvation" conducted by the U.S. against the Japanese during the latter part of the war in the Pacific. U.S. aircraft laid over 12,000 mines in Japanese harbor approaches and shipping routes. Altogether, 650 Japanese ships were sunk by these mines and maritime shipping was totally disrupted. From India, Australia, Hawaii, and hard-won island bases, Allied ships, submarines and aircraft hunted down and attacked Japanese shipping. Mines were sowed in the shipping lanes along the Asian coast from Rangoon to Korea and Japan. Mines were even laid in the interior rivers of China. In the final stage, U.S. Air Force B-29's saturated Japanese home waters with mines. The Japanese, being very dependent on imports, nearly ceased to exist as a maritime force. (See Figure 1.5)

The Japanese Navy was not prepared to cope with the influence mines which were planted in their home waters by the U.S. Those ships not sunk by mines were either forced to stay in closed ports or divert to a few overcrowded ports where they were subject to aircraft and submarine attack.

Ground mines rely on three "influences" for actuation: residual magnetism in a steel hull, noise and pressure. Magnetic influence was invented by the Royal Navy in the First World War but used on a large scale by the German Navy and the German Luftwaffe during the Second World War. Acoustic influence followed rapidly and by 1943, the first "oyster" or pressure-actuated mines were in use. This last type remained intractable and, by 1945 were generally conceded to be almost unsweepable. Fortunately, all influence mines suffer from the inherent limitation of requiring shallow water and

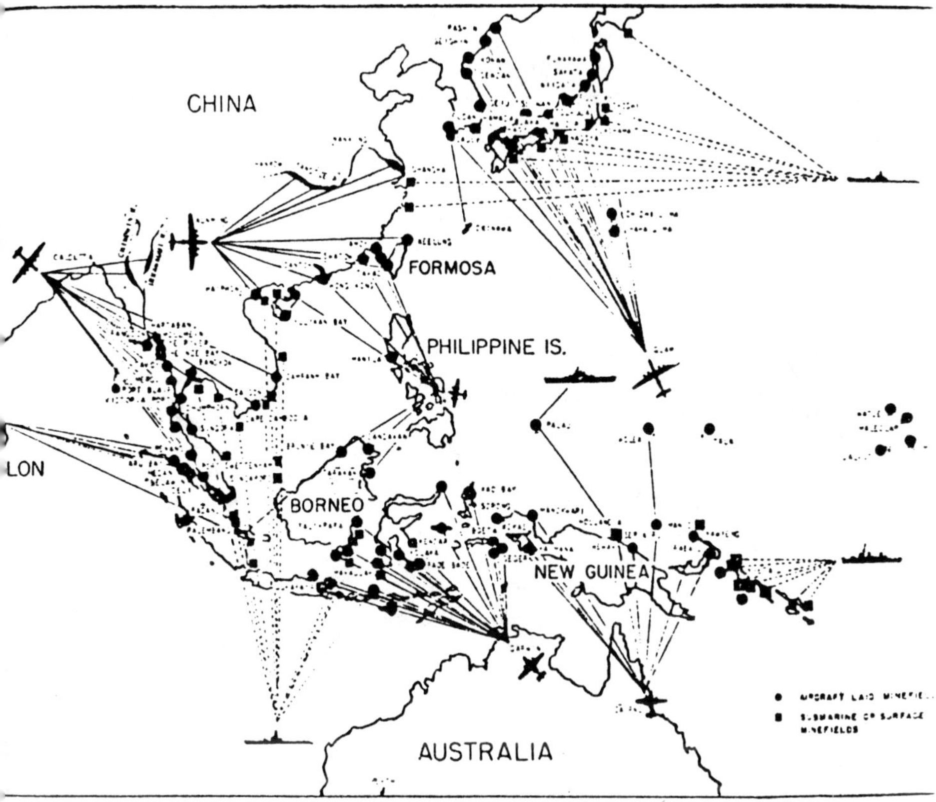

Figure 1.5. Offensive mine campaign against Japan showing operating bases and areas of coverage 1942–1945.

proximity to the target to be effective. The oyster mine was, in fact, designed to protect invasion beaches and relied on the pressure-wave set up by large deep-draft hulls moving through the water. There is also the problem of handling and safety if actuation is made too sensitive.

During World War II, for every 35 mines laid, one ship was sunk. This shows the cost effectiveness of mine warfare. The Europeans with their long sea frontiers and easily mined shallow seas have become true experts in sea mine warfare during the course of two World Wars. During World War II alone the Germans laid about 150,000 mines; the British close to 200,000. The damage done to Allied and Axis shipping was about even, with each losing over 1000 ships to mines. The strategic importance of mine warfare can be gathered from the fact that in 1941 the Soviet Fleet was not able to penetrate the mine fields the Germans had placed across the whole width of the Baltic. No major

Soviet ship based at Leningrad succeeded in reaching the open sea as long as the German Navy was able to actively defend the barrier.

The minesweeping and clearing effort of the mines laid starting in 1939 during World War II has continued for many years, and has not been completed. It will not be until the last mine has rusted or been rendered useless. These mines were rather unsophisticated devices which with the sweeping means now available, could be easily cleared, if their triggering circuits were not corroded or even destroyed with the passage of time. The explosive charge, however, often remains intact, making such old mines extremely dangerous, as their responses are totally unpredictable.

In World War II, Allied mines were notorious for doing substantial harm to Allied forces, especially where defensive mine fields were deployed. In many instances, the accidents were caused by lack of caution on the part of commanding officers and masters.

An example of this occurred in the channel of Tutuba at the southeast corner of the island of Espiritu Santo in the New Hebrides Islands, code name Base Button, in January of 1943. Both ends of the channel were mined. It was a mandatory procedure for all Navy and civilian ships, supporting the war effort to stop, before entering the harbor, to pick up a pilot who knew the path through the minefield.

Early one evening a troopship appeared, but—instead of waiting for a pilot—proceeded into the minefield and the expected happened. With a gaping hole in her side, the master headed the ship to the shore to beach her. But being a beach with a sharp drop-off, the ship sank in about 18 meters of water, with the bow shallowest. From the air it could be easily seen. Miraculously, everyone got off. The biggest loss, scuttlebut had it, was a cargo of beer and hard liquor. No report of salvage efforts was made.

As in past wars, peacetime brought the need to clear the many mines that still endangered shipping throughout the world. These clearing operations were long, dangerous, and expensive. The problem remained of drifting mines, 40,000 of which were estimated to be loose in the Pacific in 1946. In WWII, U.S. mines sank more Japanese ships than the U.S. submarine force.

MINES IN THE KOREAN WAR

At the end of World War II the U.S. Navy stood unchallenged as the world's preeminent sea power. Less than five years later a U.S. commander reported, "The U.S. Navy has lost control of the sea." A fifth rate North Korean communist power had mined the Korean harbor of Wonsan. Offshore, more than a hundred U.N. ships in a Task Force with thousands of embarked troops and equipment steamed for a week in the tedious circles of "Ooperation Yo-Yo." Minesweepers (seven of which sank) struggled to clear a channel through the 3,000 mines the North Koreans had laid.

During the Korean war 50,000 UN troops waited for eight days to land at Wonsan on the eastern coast of Korea. It had been discovered that a makeshift North Korean minelaying force, made up largely of junks, had laid a field of 3500 Soviet-supplied magnetic mines. Not even wooden-hulled minesweepers could avoid being blown up, for the Soviet mine actuators were sensitive enough to be triggered by the magnetism of engines and hull fittings. Wonson sent a shock through Western navies, who had all forgotten that mine warfare was the only field in which the Russians had consistently shown competence. As long ago as 1855, Russian mines had inflicted damage on enemy ships, and their mine warfare specialists had repeatedly shown skill and ingenuity.

An equally bad situation almost existed at Inchon on the western coast of Korea with the U.N. forces there. North Korean forces assisted by the Soviet Union, were within a day or two of mining Inchon when the amphibious landing took place. During the first two years of the Korean War, all Navy ship losses and 70% of Navy casualties could be attributed to mines.

As a result of the bad Korean experience, the U.S. Navy quickly restored mine warfare capabilities. Layed-up sweepers and new ships entered the fleet. The ranks of the mine warfare force increased. Every fleet training exercise had to include a mine warfare phase. The helicopter emerged as a capable countermeasures vehicle. It seemed that finally the U.S. Navy had learned the lessons of the 20th century. Unfortunately, these lessons were soon forgotten again.

MINES IN VIETNAM

By the time the Vietnam war came along, mine warfare had expanded beyond the scope of its sensor and weapon systems. Triggering had grown to include not only contact and magnetic fuses, but sound and pressure as well. Admiral Moorer, the Chairman of the Joint Chiefs of Staff, reported in 1972 that the "handwringers" in the Executive Branch were fearful that the Soviets would immediately rush their minesweepers to Haiphong Harbor in North Vietnam if it were mined and sweep them in in a matter of hours. This was a ridiculous overestimate of the enemy situation, and capability.

President Nixon, after being told that a mining plan was in existence, ordered the harbor mined. On the morning of the operation, it took one aircraft carrier and 26 aircraft one hour to mine the Haiphong Harbor. Not one ship entered the harbor until the end of the war, when U.S. minesweepers cleared the channel. Up to the mining of the harbor, the peace talks in Paris were not making any progress. After the mining, the North Vietnamese delegation began deliberate negotiations quickly.

A new family of mines called "Destructors," a mine made from a bomb, first came into use in 1967 during the Vietnam Conflict. The term destructor was used to reduce the political ramifications from the use of the term "mine." These bomb/mines contained sophisticated electronic mine firing mechanisms that were inserted in the fuse cavities in the bombs. Early versions of the mechanism were magnetic, but by the end of the mining campaign these destructors contained magnetic and seismic mine firing mechanisms.

Some inland waterways were mined by the U.S. Navy in the Vietnam area of operations.

MINES IN THE MALVINAS-BRITISH WAR (THE FALKLANDS)

Horned mines, as the name suggests, are operated by electro-chemical or simple electrical switches protruding from a buoyant casing. Locally made copies of a German Second World War type were found in Port Stanley in the Malvinas (Falklands) during the Argentine-British fighting in 1982. Royal Navy mine countermeasures ships wer used to clear them after the Argentine surrender in June 1982. The British were expec ting to find some of their own and other influence mines (known to be in the Argentin inventory) and were relieved to have a comparatively simple task.

Great Britain used converted fishing trawlers to clear Argentine mines in the Malvina operations.

MINES IN THE MIDDLE EAST CONFLICT

Mine warfare never seems to get the attention and priority it deserves until a crisis arrives that dramatizes the seriousness of the threat. It should be said however, that the Europeans and the Soviets do place a reasonably high priority on mine warfare. This is evinced by the funding that they put into new vessels and technology for mines and mine countermeasures.

In 1956 in a Suez incident, Egypt stated that the approaches to Alexandria had been mined and that the Gulf of Suez was dangerous to navigation. This discouraged ships from entering the area.

In 1967 during the Israeli 6-day war, Egypt announced a blockade of the Gulf of Aqaba, stating that mines had been laid. Again this discouraged shipping.

The Soviets did mine sweeping during the 1967 to 1973 Arab-Israeli wars. Both sides laid mines. Some Soviet mines were found in later years that had been laid during the Arab-Israeli conflicts.

Even the intervening years saw the use of mines. Mining and other blockages – material and political – closed the Suez Canal for years, forcing expensive changes to the world shipping patterns.

In the 1974 Mideast war, mines were used in the Suez Canal and its approaches. They were cleared by U.S. helicopters towing MK-105 sleds carrying detonating systems. Later the same year, mysterious mines of unknown origin appeared in the Gulf of Suez. Again mine clearing was ordered, this time a multinational effort, with the U.S. using the sleds to complement the operations of ships and salvage ships, whose crews included divers.

In 1984 "Mystery" mines appeared in the Red Sea and the Gulf of Suez. Terrorists groups claimed responsibility, but evidence pointed to Colonel Gadhafi of Lybia as the perpetrator and the Soviet Union as his supplier.

Horned mines were also laid by the Iranians in the Gulf, and have proved just as effective as modern mines in disrupting shipping movements in the region.

In July 1984, several merchant ships proceeding to and from the southern end of the Suez Canal were damaged by mysterious underwater explosions. In response to an urgent appeal by the Egyptian government, an international force of U.S. Navy helicopters and Italian, French, British and Dutch minehunters was assembled quickly and deployed to the area. The culprits turned out to be a new type of Russian influence mine. These mines that were planted by the Libyans were meant to sabotage Egyptian control of the Canal without causing deaths and thereby incurring the risk of military intervention or retaliation. (See Figure 1.6)

On 12 September 1985 the British minehunter *Gavinton* detected a mine-like object

Figure 1.6. Russian mine retrieved by Royal Navy south of Suez Canal, Egypt on 12 September 1985.

on the bottom in 150 feet of water, about 15 miles south of the entrance to the Suez Canal near Port Tewfik. A Swedish civilian "Sea Owl" remote-controlled submersible had been used for several days by the Royal Navy to examine the mine where it lay. After determining that it was safe, the mine was moved to shallow water to be inspected by divers and cut into pieces. The detonator/instrumentation section was brought on board the *Gavinton* for inspection and later sent to London for extensive analysis.

This mine was unknown in the West. Even *Jane's Weapons Systems* had no data on it. It was 10 feet long and 21 inches in diameter and loaded with about 400 pounds of high explosives, although capable of holding over 1,500 pounds. It was apparently the intent of the country that laid the mine in the Red Sea to scare and only do slight damage to shipping rather than sink ships.

The mine had not been in the water very long since its casing was clean. It was designed to rest on the bottom after being laid by a submarine or surface ship. There was no indication that it could be laid by aircraft. The 21 inch diameter of the mine showed that it could be torpedo-tube launched.

The mine had an acoustic, pressure and magnetic detonator and was marked "99501 NG63." According to Egypt's Defense Minister, it had been set to be armed on or about 27 July 1985, but a malfunction in the arming mechanism rendered it harmless.

The mine was Soviet in origin, designed for export, and manufactured no earlier than 1981 since serial numbers on both launching rail lugs ended in 81. Although the mine was of Soviet manufacture, there was no reason to assume that the Soviets were directly responsible for the mine's presence in the Gulf of Suez. The Russians said that they were not in any way connected to the minelaying, which they strongly condemned.

The evidence is fairly conclusive that the Libyan ship *Ghat* had laid a number of sophisticated Soviet "export" mines which had created the "Mines of August Crisis." This showed how easily Third World nations could mine and how difficult it is to bring the culprit to justice. How many more events such as this will occur in the future? The world has seen that the disruption of world commerce can occur by such an operation.

In mid-1987 there was yet another "Mines" crisis, again centered in the same Mideast region: this time in the Persian Gulf. After the reflagged tanker *Bridgeton* struck a mine on 24 July 1987, the U.S. Secretary of Defense, Caspar Weinberger, stated that "we were not looking for mines there because we had never seen a mine in that area." This comment was made despite the fact that at least four other ships had previously detonated mines in the area in little more than two months since the Iraqi attack on the *USS Stark* (FFG-31). This showed the world once again that the U.S. gave low priority to mine warfare. The scene viewed on TV sets the world over was of three U.S. Navy warships being let through suspected minefields by the damaged tanker *Bridgeton*, which the Navy's combatants were supposed to be protecting. This hardly helped the image of the U.S. "Superpower."

The Iranians used small, fast boats to lay a relatively small number of moored mines in the path of the first U.S. convoy through the Gulf. They declared certain areas as restricted which restricted neutral shipping to a channel about two miles wide. This was much easier for Iran to mine with their limited numbers of mines and increased the chances that vessels using the channel route would encounter a mine.

Deliberately floating mines are banned by international law because their movements cannot be predicted and they may attack neutral shipping outside a war zone. Ironically, the first victim of the Iranian mining effort was a vessel carrying Iranian oil.

In August 1987 the U.S. Navy caught an Iranian landing craft, the "Iran AJR" red-handed laying mines in the Gulf in attempts by Iran to interrupt oil shipments from Iraq's Kuwati allies. (See Figure 1.7) The horned contact mines were believed to be North Korean, supplied originally by the Soviet Union. They were MO8 type tethered contact of World War II design.

Despite massive U.S. retaliation against Iran, this did not prevent Iranian mines from damaging the *USS Samuel B. Roberts*, an FFG-7 type frigate in 1988. The remarkable humiliation of the U.S. Navy in the Gulf in 1987 showed how far its MCM capability had slipped. The British, French, and Germans had already invested large sums of

Figure 1.7. MO8 Type Contact Mines on deck of Iranian AJR ship in Persian Gulf September 1987.

money in new mine ships. However, the U.S. Navy's MCM ships were so long delayed that they could muster only three 30-year-old minehunters. New ships were under construction and even in service but considerable time was needed to train operators on new equipment. Minesweepers from Great Britain, France, Belgium, Italy, The Netherlands and the U.S. were sent to the Gulf.

CHAPTER 2

Minelaying, Minesweeping and Mine Countermeasure Platforms

MINE WARFARE VESSELS/AIRCRAFT

World Navies

Algeria

2 of T-43 Type—Soviet Union. 500 tons, 14 kts.

Argentina

6 of UK "Ton" class. 370 tons, 15 kts. Plessey 193M sonar.

Australia *(see Figure 2.1)**

2 MH CAT catamaran minesweepers—Fiberglass hulls. 100 tons, 10 kts. Equipped with portable MWS-80 KAE German sonar.
3 of UK "Ton" class coastal minesweepers.

Belgium *(see Figure 2.2)*

10 Tripartite minesweepers. 511 tons, 15 kts. Bldr: Polyship Consortium. Glass-reinforced plastic hulls. Two PAP-104 remote-control mine locators. Sonar: French DUBM-21A.
7 AM/MSO oceangoing minesweepers. 720 tons, 15 kts. SQQ-14 sonar.
6 US/MSC coastal minesweepers. Two with Plessey 193M sonar, four with UQS-1 sonar.
13 Herstal class inshore minesweepers. 160 tons, 15 kts.
2 Mine countermeasure support ships.

Brazil *(see Figure 2.3)*

6 German Schutze class R55 minesweepers. 230 tons, 24 kts.

* Note: Figures and Tables are at the end of this chapter.

Bulgaria

4 Soviet Vanya class minesweepers. 220 tons, 18 kts.
2 Soviet T-43 class minesweepers. 500 tons, 14 kts.
4 T-301 class minesweepers. 146 tons, 12 kts.
1 Soviet Yevgenya class inshore minesweeper. 80 tons, 12 kts. Plastic hull. Has towed television minehunting and marking system. Effective to depths of 30 meters.
12 PO-2 class inshore minesweepers. 50 tons, 9 kts.

Canada

6 Bay Class. Formerly minesweepers—reclassified as Reserve Patrol Escorts. Various CL-28 (CP-121) Argus Maritime Patrol 4-engine aircraft can carry mines. CP-140 Aurora also can be so configured. Using small portable sonars, a number of smaller Canadian ships can be outfitted as minesweepers/hunters. Study is being done for design and building of its own Maritime Coastal Vessel Defense (MCDV). May combine minesweeping and patrol capabilities.

Chile

A number of small boats could be configured for minelaying or minesweeping duties.

China—PRC

20 Soviet T-43 class fleet minesweepers. 610 tons, 14 kts.
100 auxiliary minesweepers convertible from fishing boats or motor junks.
14 ex-U.S. LSM-1 class—some used as minelayers.
— Type 312 drone minesweeper. 21 mtr. hull. Radio-controlled to 3 NM range.

Cuba

2 Soviet Yevgenya class inshore minesweepers. 80 tons, 12 kts.

Denmark

10 Willemoes class GM boats. Can be equipped to lay 20 mines in lieu of torpedo tubes. 230 tons, 40 kts.
4 Falster class minelayers. 1880 tons, 16 kts.
2 Lindormen class minelayers. 575 tons, 14 kts.
8 Ex-U.S. coastal minesweepers. 350 tons, 13 kts. UQS-1 sonar.

Dominican Republic

2 Ex-U.S. Admirable class minesweeper-escorts. 600 tons, 15 kts.

Egypt

6 Soviet Romero and 6 Soviet Whiskey submarines. Can carry 28 mines or 14 torpedoes.
4 Soviet Skory class destroyers. Can carry 50 mines.
12 Soviet SO-I class patrol boats—can carry mines. 80 tons, 40 kts.

4 Ex-Soviet Yurka class minesweepers. 460 tons, 18 kts.
6 Ex-Soviet T-43 class minesweepers. 500 tons, 14 kts.
2 Soviet T-301 class minesweepers. 145 tons, 12 kts.

Finland *(see Figure 2.4)*

2 Minelayers. 1100 tons, 20 kts.
6 Kuha 22 class inshore minesweepers. 90 tons, 12 kts. GRP hulls with some stainless steel and aluminum. The Kuha minesweeper can control the 20-ton Kiiski type drones.

France *(see Table 1 and Figures 2.5 through 2.9)*

A number of ex-Canadian, U.K. and French minesweepers have been converted to perform other duties.
15 Eridan class minehunters. 500 tons, 15 kts. DUBM-21B sonar and EVEC 20 automatic plotting table. Two each PAP 104 remote control mine locators. Hull built of glass and polyester resin. One mechanical drag sweep. Common design with France, Belgium and Netherlands.
5 Circe class minehunters. 460 tons, 15 kts. DUBM-20 sonar. Hulls of laminated wood. Designed for detection and destruction of mines as deep as 60 meters. Mines can be destroyed by PAP-104 (Poisson auto-propulse) wire-guided underwater sled device, 217 mtrs. long, 1.1 mtrs. in dia, weighing 700 kg. Moved by two electric motors that drive it at 6 knots for a distance up to 500 mtrs. Has a TV camera that displays an image of the mine. Can deposit an explosive charge of 100 kg. near mine, move away and charge is then detonated by ultra-sonic waves.
5 Ex-U.S. Agile class converted to minehunters. 700 tons, 14 kts. Two PAP-104. DUBM-21 sonar with EVEC 11 automatic plotting table.
5 Ex-U.S. Agile class ocean minesweepers. 700 tons, 14 kts. French "Dompaire" class. Wooden hulls. Some converted with DUBM-41 sonar. Some had AP-4 drag sweep. Newer DUBM-42 sonar.
5 Coastal Sirius class minesweepers. 400 tons, 15 kts. French-built versions of British ton class.
9 Ex-U.S. Adjutant class. 300 tons, 13 kts.

A number of French Atlantique ASW aircraft and other aircraft such as attack type that can carry larger bombs can be configured to lay mines.
10 Planned BAMO (Batiments Antimines Oceaniques). Minehunting to 300 mtrs. depth. May be 800-ton catamaran of GRP.

Germany GDR

22 (1969*) Kondor I class inshore minesweepers. 225 tons, 17 kts. Mechanical, acoustic and remote-controlled sweeping gear. Equipped with sonar.
28 (1972*) Kondor II class inshore minesweepers. 310 tons, 17 kts. Mechanical, acoustic and remote-controlled sweeping gear 1-Ss/E. Has sonar.

*Date of entry into fleet.

Note: Most of the mine defense vessels of the Warsaw Pact navies can take on mir as on-deck cargo and deploy them as barriers. The vessels have shifting trac on the side decks and launch ramps on the stern. Offensive mine barriers mission of fast surface ships, submarines and naval aviation.

Germany FDR *(see Figures 2.10 through 2.17)*

Type 355 Pressure-mine sweepers. A project to design a ship capable of detecti and disposing of the "unsweepable" pressure mine. Planned for the 1980s.

10 Type 343 minelayer/minehunters. To use DSQ-11 minehunting sonar. 590 tor steel hull, non-magnetic.

6 Type 351 drone minesweeper-control ships. 465 tons, 16 kts. Type 3: minesweepers to be converted in 1980s. Each ship to control three F-1 "Troik drone magnetic mechanical minesweepers.

18 Type F-1 Troika drones. 100 tons, 9 kts. Magnetic and mechanical sweep.

10 331A minehunters. 420 tons, 17 kts. DSQS-11 sonar by KAE. Wooden hulls. Car two French PAP-104 remote minehunters.

2 331B minehunters. Have British Plessey 193M Mk 20G sonar. No PAP-10 "Lindau" class.

22 Type 340A and Type 341B minesweepers. 240 tons, 24 knots. Some are minelaye and minesweepers.

1 Type Holnis inshore minesweeper trials ship for Troika magnetic minesweepi system. Troika uses three radio-controlled, 27 mtr. Seekuh (Sea-cow) launch

18 Types 393A and 394B inshore minesweepers. 185 tons, 14 kts. Wooden hul

21 Type Schutze coastal minesweepers. 305 tons. Type MW-V 50 mi countermeasure vessel. Minehunting, minesweeping and minelaying. No magnetic steel hulls.

10 Type 332 fast minesweepers. 400 tons. Non-magnetic steel hulls. DSQS-11 son: two Penguin B3 remote vehicles.

Great Britain *(see Figures 2.18 through 2.26)*

12 MSM/EDATS—Extra Deep Armed Team Sweeps. Adaptation of commerci stern-haul trawlers. To sweep deep-ocean mines by operating in pairs, with wire sweep between them. In time of war, additional commercial fishing cra would be acquired for the same duties.

25 Hunt class minehunters. 615 tons, 17 kts. Hulls of glass-reinforced plastic. Car two French PAP-104 wire-guided, remote-controlled locators. Acoustic ar magnetic sweeps. Builder: Vosper Thornycroft (UK) Ltd.

— Sandown class SRMH (Single Role Mine Hunter). With Plessey 2093 sonar. Tł sonar transducer is lowered beneath keel. Uses PAP-104 MK 5 built by EC Honeywell. Builder: Vosper Thornycroft. 450 tons, 15 kts.

33 "Ton" class minehunters. 370 tons, 15 kts. The survivors of a class of 118 con pleted in 1952-58. Use type 193 sonar. Wooden hulls, sheathed with nylon belc the waterline. Use fin stabilizers. M-1141 carries the Osbourne multiple-swe array. M-1182 was given pump-jet propulsion.

— COOP (Craft of Opportunity) vessels could be outfitted with remote mine dispos vehicles such as ARMS—Plessey Ltd. Advanced Remote Minehunting Systen a remote-operated vehicle to 300 meters. ARMS has its own sonar and a cable to the ship.

1 Mine countermeasure support ship. 1375 tons, 16 kts.

— Mine countermeasures Hovercraft. 62 tons, 60 kts. Bldg. Vosper. Surface Effects Ship, or Hovercraft. British Hovercraft Corp. has carried out trials with their SRN6 and BH7 hovercraft. The Hovercraft has low underwater acoustic and pressure signatures, good for working in a minefield. All types of sweep and minehunting equipment can be operated from them. Certain of the British aircraft such as the Nimrod maritime patrol ASW aircraft could be configured to carry mines.

Greece

15 Ex-U.S.A. Adjutant class and Falcon class minesweepers. 300 tons, 13 kts.

2 Ex-U.S.A. minelayers. 720 tons, 13 kts. Carry 100 to 300 mines, depending on type.

Guinea

1 Ex-Soviet T-58 class ocean minesweeper.

India

12 Ex-Soviet Natya class. 650 tons, 20 kts. No ramp on stern as on Soviet Navy ships. Steel hulls.

12 Ex-Soviet Yevgenya class. 90-ton inshore minesweeper. Small dipping sonar aft.

4 Ex-British "Ton" class minesweepers. 370 tons, 15 kts.

4 Ex-British "Ham" class minesweepers. 120 tons, 14 kts. Some built by Indian yards have teakwood hulls.

Indonesia

5 Ex-Soviet T-43 class minesweepers. 500 tons, 14 kts.

2 German Raum Boote class inshore minesweepers.

— Netherlands Alkmarr Tripartite design is being purchased.

Iran

3 Ex-U.S.A. Falcon class minesweepers. 320 tons, 12 kts.

2 Ex-U.S.A. Cape class inshore minesweepers. 180 tons, 13 kts.

Iraq

2 Soviet T-43 class fleet minesweepers. 500 tons, 14 kts.

3 Soviet Yevgenya class inshore minesweepers. 80 tons, 12 kts.

Ireland

3 Ex-British "Ton" class minesweepers. 370 tons, 15 kts.

Italy *(see Figure 2.15)*

4 Lerici class minehunter/minesweepers. 485 tons, 15 kts. Bldr: Intermarine La Spezia. Glass-reinforced, shock-resistant plastic construction.* Can carry two PAP-104 minehunting devices and conventional sweep gear. SQQ-14 HF sonar with retractable transducer. Also sold to U.S.A. under license for construction there as Lerici II, but 2 meters longer. GRP hull is molded as single unit. 140 mm. thick, very heavy.

4 Ex-U.S. Agile class fleet minesweepers. 665 tons, 14 kts. Sonar UQS-1.

30 U.S.A. Adjutant class minesweepers. 375 tons, 13 kts. Sonar UQS-1. Wooden hulls and antimagnetic fittings.

10 Aragosta class inshore minesweepers. 120 tons, 14 kts. Based on British "Ham" class. Wooden hulls.

Japan *(see Figures 2.30 and 2.31)*

18 Aerial minelaying versions of the C-130 Hercules aircraft.

1 Soya class minelayer. 2150 tons, 18 kts. Has platform for minesweeping helicopter.

1 Hayase class mine countermeasures support ship/minelayer. 2000 tons 18 kts. Carry 200 miles.

17 Hatsushima class minehunter/minesweepers. 440 tons, 14 kts. Equipped with Type 54 mobile minehunting devices, which carry and lay their own disposal charges.

19 Takami class minehunter/minesweepers. 380 tons, 14 kts. ZQS-2 sonar, built under license from British Type 193M. Wooden hull, some GRP.

11 Kasado class minehunter/minesweepers. 330 tons, 14 kts. ZQS-2 sonar. Wooden construction. Earlier ZQS-1 was Japanese version of UQS-1.

6 Inshore minesweepers. 58 tons, 10 kts.

— Oceangoing mine countermeasures vessel. 1000 ton. With Type S7 deep-sea minehunting vessel and S8 deep-sweep gear. Five are planned.

Korea—South ROK

5 U.S.A. MSC-289 class coastal minesweepers. 315 tons, 14 kts. Wooden hulls. Bldr: Peterson Builders.

3 U.S.A. MSC-268 class coastal minesweepers. 320 tons, 14 kts. Wooden hulls.

1 Ex-U.S.A. minesweeping boat. 44 tons, 12 kts.

— Swallow Craft (Kangnam Shipbuilders) building GRP vessel. Similar to Italian Lerici design. Plessey 193M sonar or later KAE MWS-80 sonar. Racal Decca MAINS information center.

Kuwait

Plan to purchase 2 Alkmaar class minehunters from Netherlands shipyard Vander Giessen-de Noord.

Libya

3 Soviet Foxtrot class submarines. 1950 tons, 16 kts. Can carry 44 mines.

*GRP/foam sandwich construction, monocoque design.

Malaysia

5 Ex-British "Ton" class coastal minesweepers. 360 tons, 15 kts. Wooden hulls over aluminum-alloy framing.

Mexico

— Has a large number of ex-U.S.A. Auk and Admirable class minesweepers. However, most minesweeping equipment has been deleted. Auk is 890 tons at 17 kts. Admirable is 650 tons at 15 kts.

Morocco

1 French Sirius class. 365 tons, 15 kts. Wooden hulls on aluminum frames.

Netherlands *(see Figure 2.32)*

15 Alkmaar class minehunters (Tripartite design). 511 tons, 15 kts. French DUBM 21A sonar, Decca HiFix precision navigation system. Hull of glass and polyester resin. Can tow mechanical sweep.

18 Dokkum class minehunters/minesweepers. 373 tons, 14 kts. Sonar 193M on minehunters only. Similar to French Sirius and British "Ton" class. Minehunters are equipped with ARNAS system that displays dropped marker buoys on the radar screen.

16 Van Straelen class inshore minesweepers. 151 tons, 13 kts. Wooden hulls. There are long-term plans to build some larger deep-sea sweepers in the late 90s. Possibly, the Norwegian air-cushion MCMV design is one of the contenders.

Norway *(see Figure 2.33)*

2 Vidar class minelayers. 1500 tons, 15 kts. Carries 320 mines.

1 Inshore mine-planter. 282 tons, 9 kts.

1 Prototype Swedish Gasten class minesweeper. 120 tons, 11 kts. Plastic and fiberglass ship.

10 U.S.A. Falcon class coastal minesweepers. 300 tons, 13 kts. Sonar UQS-1. Some converted to British type 193M sonar and two with French PAP-104 remote-controlled minehunting device.

10 In 1987, planned to build 10 SES (Surface Effects Ship) class coastal minesweeper/minehunters. Called Sauda class.

Paskistan

7 Ex-U.S.A. coastal minesweepers. 320 tons, 13 kts. UQS-1D sonar. Wooden hulls. Similar to U.S. Falcon class.

Philippines

2 Ex-U.S.A. Falcon class coastal minesweepers. 320 tons, 13 kts.

Poland

4 Ex-Soviet Whiskey class submarines. 1050 tons, 16 kts. Can carry 28 mines.
12 Krogulec class minesweepers. 450 tons, 16 kts. Mechanical sweeping gear.
12 Soviet T-43 type minesweepers. 500 tons, 14 kts.
28 K-8 class inshore minesweepers. 26 tons, 12 kts. Wooden hulls. Wire sweeps only.
2 Notec class coastal minesweepers. 250 tons. GRP hulls. Mech. and Acoustic sweeping gear.

Portugal

4 Sao Roque class minesweepers. 400 tons, 15 kts. Wooden hulls, fin stabilizers. Similar in appearance to British "Ton" class.

Rumania

4 Democratis class minesweepers (German M-40 class). Bldr: Galati 1943-1951. 653 tons, 17 kts.
18 Ex-Soviet T-301 class minesweepers. 145 tons, 12 kts.
8 Polish TR-40 class inshore minesweepers. 70 tons, 18 kts.

Saudi Arabia

4 U.S. MSC class mine warfare ships. 320 tons, 14 kts. Bldr: Peterson.
– Plan to purchase Vosper Thorneycroft, Southampton, U.K. Sandown class minehunters. Saudis leasing two 615-ton Hunt class ships in interim.

Singapore

2 Ex-U.S.A. Bluebird class minesweepers. 300 tons, 13 kts. Sonar UQS-1D.

South Africa

4 Ex British "Ton" class former minesweepers. Retain most sweep gear and could be used to sweep moored and mechanical mines.
6 Ex-British "Ton" class minesweepers and minehunters. 370 tons, 15 kts. Some have British 193M sonar, and PAP-104 remote-controlled minehunting devices.

Spain

4 Ex-U.S.A. Aggressive class minesweepers. 665 tons, 14 kts. Sonar SQQ-14. Equipped with mechanical, magnetic and acoustic sweeping.
11 Ex-U.S.A. Adjutant MSC-268 and Redwing class minesweepers. 355 tons, 13 kts. UQS-1D sonar. Spanish Navy is presently involved in choice of a successor design and may modernize some of present ships to keep them operational.

Sweden *(see Figures 2.34 through 2.38)*

1 Minelayer Karlskrona. 3100 tons, 20 kts.
2 Alvsborg class minelayers. 2660 tons, 16 kts. Can carry a helicopter.
1 Converted merchant ship minelayer. 4250 tons, 14 kts.

12 (1958) Arko class coastal minesweepers. 285 tons, 14 kts. Wooden hulls.
3 Gassten class inshore minesweepers. 120 tons, 11 kts. One of these had a glass-reinforced plastic hull. Intended to serve as a prototype of a new class of 300-ton, 43-meter minesweepers.
7 Hisingen class inshore minesweepers. 140 tons, 9 kts. Wooden hulls.
8 M-15 class inshore minesweepers. 70 tons, 12 kts. Wooden hulls.
6 Landsort class MCMW. With RACAL MAINS minehunting system. GRP hulls. Thomson Sintra TSM2202 MH sonar. Control 3 SAM drone sweeper catamarans.
– Type M70 minehunter GRP. 300 tons, 14 kts.
1 Alvsborg minelayer and depot ship. 2660 tons.

Syria

1 Soviet T-43 class fleet minesweeper. 500 tons, 14 kts.
2 Ex-Soviet Vanya class coastal minesweepers. 200 tons, 18 kts.

Taiwan

14 Ex-U.S.A. and ex-Belgian Adjutant MSC-268 and MSC-289 classes of coastal minesweepers. 320 tons, 12 kts. UQS-1D sonar. Wooden hulls.
1 Ex-U.S.A. minesweeping boat. 40 tons, 12 kts. Wooden hull.
8 Ex-U.S.A. minesweeping launches. 24 tons, 8 kts. Wooden hulls.

Tanzania

2 East German Schwalbe class former inshore minesweepers.

Thailand *(see Figures 2.39 and 2.40)*

2 Bangrachan class minelayers. 319 tons, 12 kts.
1 Mine countermeasures support ship. 586 tons, 12 kts.
4 Ex-U.S.A. MSC-289 class coastal minesweepers. 330 tons, 13 kts. UQS-1D sonar. Wooden hulls.
5 Ex-U.S.A. motor launch minesweepers. 21 tons, 8 kts. Wooden hulls.
2 M48 MCMV. Bldr: Lurssen—W. German Lindau class. Wooden hulls. 448 tons. W. German KAE sonar. Two Italian Pluto Mine Disposal Vehicles.

Turkey

1 Danish Flaster class minelayer. 1880 tons, 16 kts.
5 Minelayers. Bldr: Brown, Houston, Texas, U.S.A. 743 tons, 12 kts.
1 Mineplanter. Bldr: Higgins, New Orleans, LA, U.S.A. 540 tons, 10 kts.
5 Ex-German, French Mercure class coastal minesweepers. 540 tons, 14 kts. Wooden hulls.
12 Ex-U.S.A. Adjutant, MSC-268 and MSC-294 class coastal minesweepers. 300 tons, 14 kts.
4 Ex-Canadian Bay class coastal minesweepers. 390 tons, 16 kts. Wood hulls on steel frames.
4 Ex-U.S.A. Cape class inshore minesweepers. 203 tons, 12 kts. Wooden hulls.

U.S.A. *(see Figures 2.41 through 2.52)*

Note: Surface minesweepers in the U.S. Navy are named for birds and terms indicating action or aggressiveness.

Surface ships, submarine and aircraft are or can be configured for minelaying.

4 MCM class oceangoing minehunters. Several already constructed; eleven authorized to 1988, possibly more later fiscal years. Under construction at Peterson Builders Inc. and Marinette Co. in Wisconsin. Wooden hulls with plastic covering. Approximately 1250 tons, 14 kts. Known as "Avenger" class. First ships use SSQ-14/30 sonars. Later to use SQQ-32 Raytheon/Thomson Co. sonar. Honeywell Co. SQQ-48 MNS (Mine Neutralization System) remote-controlled minehunting system on board.

1 MHC-51 "Lerici" class minehunter-coastal. PINS (Precise Integrated Navigation System) on board the MSM-1 class permits the ship to maneuver through minefields. Intermarine Co. of Italy under license to assist U.S.A. firm. Expected to enter service in 1991. Funding for the second and third of a planned force of 17 of these ships was included in the FY 89 budget request, but some may not be authorized due to budget constraints. 760-ton mine countermeasure vessel. Uses GRP (Glass-Reinforced Plastic) monocoupe construction. Hercules Co. of Wilmington, Delaware joined with Intermarine of Italy to form Intermarine U.S.A. on this project. The vessels are scheduled for lamination at the Sayler Marine facility in Savannah, Georgia.

1 MSH-1 Cardinal class. This program has been cancelled by the U.S. Navy. Mention is made of it here only to cover past events. Had twin hull with rigid sidewalls, derived from SES (Surface Effects Ship) designs. Used GRP foam-sandwich hull construction. Built by Bell Aerospace Textron Halter yard at New Orleans, Louisiana. Tests at sea not successful due to explosions of mines having adverse effects on rigid sidewalls of the vessel. A sad commentary on the U.S. Navy's ability to purchase mine warfare vessels.

21 MSO class oceangoing minesweepers. Most in U.S. Navy Reserve Fleet. About 30 years old. 665 tons, 14 kts. Wooden hull construction. Non-magnetic stainless steel machinery. Have SQQ-14 General Electric sonar. Some have SSN-2 precise navigation system. Were of "Aggressive" and "Dash" classes. Woods used: Alaskan cedar, Douglas fir and white oak.

2 Acme class oceangoing minesweepers. 720 tons, 14 kts. Similar to "Aggressive" and "Dash" class, but slightly larger. Originally equipped as Mine Division flagships.

Note: The U.S. Navy built 58 wooden minesweepers during the Korean War. Twenty were lost and 35 were sold to allied navies. Only three remain active. Most U.S. Navy minesweepers are in the Naval Reserve Force.

1 MSB-29 class minesweeping boat. 80 tons, 12 kts.

7 MSB-5 class minesweeping boats. 30 tons, 12 kts. Survivors of a class of 47 built between 1952 and 1956. Wooden hulls, non-magnetic machinery (including two sweep generators).

– COOP (Craft of Opportunity Program). A plan to fit smaller ships out for limited minesweeping duty. An example of this is the U.S. Navy YP-676 class yard patrol craft. Other COOP vessels could be small fishing vessels, Naval launches, Dhows, or Coast Guard vessels. The British Navy has a similar program. Modular minehunting equipment including position-fixing gear and remotely operated vehicles could be installed when needed. Manned by Naval Reserve crews. Inexpensive side-scan sonar equipment is being procured. In an emergency, additional ships could be taken from commercial assets and outfitted with off-the-shelf navigation and sonar equipment. Twenty-two U.S. East Coast and Gulf Coast COOP units are planned. However, in a fast-reaction type of conflict, obtaining and outfitting additional COOP units, though useful in the long run, may take too long to be of value in the initial stages of NATO requirements for SLOC (Sea Lines of Commerce) clearance. Without adequate higher-level commitment, the COOP program may never reach its goals.

21 RH-53D Sea Stallion helicopters equipped with aerial minesweeping equipment. Can tow MK-103 cutters, MK-104 magnetic minesweeping arrays, MK-105 hydrofoil sled, MK-106 acoustic sweep array and SPU-1 shallow water sweep rig.

8 MH-53E helicopters are later configuration. Schedule is for 35 of these to be procured. Sikorsky Co. builds these "Sea Dragons." ALQ-166 is the new towed sweeping equipment by Westinghouse.

– U.S. Navy P3C Orion Maritime Patrol ASW aircraft. Can lay mines.

– U.S. Navy S3A Viking carrier-based ASW aircraft. Can lay mines.

– U.S. Navy A-6 and A-7 carrier-based attack aircraft. Can lay mines.

– U.S. Air Force B-52D bombers of Strategic Air Command have a secondary mission as minelaying aircraft.

– U.S. Air Force C-130H cargo aircraft. Prototypes of this aircraft have been tested as minelayers. A Gravity Drop High Volume Mine Layer (HVML) configuration was designed by Babcock Power Ltd. of England and Lockheed Co. It utilizes roll-on modular pallets. Any services that have this type aircraft could configure them in this manner.

SUBMARINES

Before World War I, the Russians had submarine minelayers. Figure 2.53 shows the Russian Krab in 1910. She did not become operational until 1915. In minelayers, the Russians were ahead of all other navies. This minelayer had been ordered in 1908. She had a horizontal mine deck with discharge ports at the stern. The mines and sinkers were stowed in cylindrical cases which were winched aft by means of a ratchet and chain, a method that was not adopted by other nations for another 20 years. (See figure 2.54.) The Russians also introduced a primitive form of snorkel to reduce the risk of being swamped while running awash.

Figure 2.55 shows the German UC submarine that was capable of launching 18 Type IV mines through six inclined mine tubes. The design was later improved to give the capability for launching 34 mines. These UC submarines were used operationally in 1915.

The British in 1915 had six E-class submarines that were converted into minelayers. These carried 32 mines in 16 tubes mounted vertically. (See Fig. 2.56.)

The U.S. Navy did not construct a mine-carrying submarine until after World War I. The Navy finally obtained authorization from Congress in 1924 to build a submarine designed to lay moored mines. The Bureau of Construction and Repair (later Bureau of Ships) designed and built the vessel, called the *Argonaut*, at the Portsmouth Navy Yard at Kittery, Maine. It was commissioned in 1928. Figures 2.57 and 2.58 show the *Argonaut* at the Washington Navy Yard in November 1928. Also shown is a side view showing how the mines were loaded and dispensed. (See Figure 2.59.) The Mine Building designed the mine assembly to fit the mine-handling equipment that was designed by the Bureau of Ships. A model of the new mine to be carried, the Mk 11, was based entirely on blueprints of the submarine. The mine and minelayer passed tests at the time so several hundred complete mines were built by the Naval Gun Factory. The *Argonaut* had storage for 60 Mk 11 anchored mines. Figure 2.4 shows a view of the *USS Argonaut* (V-4) in New York in March 1929. As can be seen, she had two six-inch deck guns. She was renamed the *Argonaut* (SS 166) in February 1931 and joined SubDiv at Pearl Harbor in 1932. On 7 December 1941, she missed being hit at Pearl Harbor because she was out on patrol off Midway Island. She was later sent to Mare Island Navy Yard for conversion to a transport submarine and carried 121 combat-loaded Marines on a raid to Makin Island in August 1942. On 10 January 1943, she was sunk by a Japanese destroyer with loss of all hands while on her third war patrol attacking a Japanese convoy off New Britain in the Southwest Pacific.

Most modern submarines lay mines by ejecting them through the torpedo tubes. The most common torpedo tubes are 21 inches in diameter, so mines must fit through this size tube. Recently the Russians have introduced 26-inch-diameter torpedo tubes on some of their submarines. These obviously could be used for larger, more powerful torpedoes or mines.

Listed here are some of the Soviet submarines and the number of mines that they can carry. (See Figures 2.60 and 2.61.)

Echo I	36	Foxtrot	44
Bravo	36	November	64
Victor	64	Zulu	44

The Yankee missile submarine—when missile tubes are removed to comply with the SALT agreement—could take an increasing load of mines.

AIRCRAFT—FIXED WING AND HELICOPTER

Many mines were laid by aircraft during World War II. U.S. Navy aircraft such as the TBF, PV-1, PB4Y-2, PBY-1, PB2Y-3, and PBY-5 were utilized first. These aircraft had only limited mine-carrying capacity. Even Royal Australian Air Force PBY Catalina Flying Boats laid mines in the Southwestern Pacific. Later, U.S. Army B-25s (2 engine), B-24s (4 engine), and then the B-29 Superfortress Bombers were utilized. The B-29 did a very effective job of mining against the Japanese in the later months of World War II.

The aircraft had become the most important type of minelayer. (See Fig. 2.62 for a view of high-speed minelaying by a U.S. Navy aircraft and Fig. 2.63 for a B-29 laying mines in Japan.) Most mining was done at night. The APQ-13 Army Radar in the B-29 bombers was a great aid in effectively planting the mines.

The U.S. Navy used carrierborne attack aircraft such as the A-6 in planting mines in North Vietnam in the 1970s. (See Figures 2.64 and 2.66 for views of the A-6 and A-7 Navy attack aircraft that are very effective minelayers.)

Figure 2.67 also shows a U.S. Air Force cargo aircraft, the C-130H, practicing dropping dummy defensive mines during tests. The U.S. Navy P-3C Series of maritime patrol aircraft can also carry a substantial load of mines, as shown in Figure 2.68. The U.S. Air Force bomber aircraft, such as the B-52, have a substantial secondary role as minelayers to supplement the U.S. Navy's efforts.

The U.S. Navy helicopters RH-53E and MH-53E have been used in the minesweeping role in various theaters of operation; both off North Vietnam and in the Middle East. (See Figures 2.69 and 2.70.)

U.S.S.R. *(see Figures 2.71 through 2.84)*

Many of the Soviet Navy surface ships, submarines and aircraft are or can be configured for minelaying. Some of their helicopters can be used for sweeping mines. Most of their minesweepers can also lay mines, many have anti-aircraft weapons on board and can perform in the ASW role.

	3	Alesha class minelayers. 2900 tons, 17 kts. Can be used as netlayers, minesweeper tenders and command ships.
* 1971	30	Natya I class fleet minesweepers. 750 tons, 20 kts. Aluminum alloy hulls. Limited ASW capability. Tamir HF sonar. Can serve as ASW escorts.
1981	4	Natya II class minesweepers/minehunters. Has stern ramp for minehunting sonar. Some used for sonar R & D, minehunting/minesweeping deep-water operations.
1964	48	Yurka class fleet minesweepers. 400 tons, 16 kts. Aluminum alloy hulls. Tamir HF sonar.
1950-60	62	T-43 class fleet minesweepers. 570 tons, 14 kts. Also Poland and P.R.C. China built under license. Over 200 were built. Steel hulls. Tamir HF sonar.
1973	50	Sonya class coastal minesweepers. 400 tons, 18 kts. Wooden hulls with fiberglass sheathing. Lightweight minehunting sonar.
1973	4	Zhenya class coastal minesweepers. 290 tons, 18 kts. GRP hulls. Not very successful design.
1963	80	Vanya class coastal minesweepers. 260 tons, 18 kts. Wooden hulls. Some used to guide Ilyusha remote vehicle drones.
1957	8	Sasha class coastal minesweepers. 280 tons, 18 kts. Steel hulls. Use Voith-Schneider propellers for easy vessel positioning.
1968	40	Yevgenya class inshore minesweepers. 80 tons, 11 kts. GRP hulls. Dipping sonar aft. Searchlight and towed TV. Minehunting and buoy-laying equipment.
1976	3	Andryusha class special minesweepers. 360 tons, 15 kts. Wooden or GRP hulls. Can sweep magnetic mines.
1976	10	Olya class minesweeping boats. 60 tons, 18 kts. GRP hulls. Mechanical minesweeping equipment.
1970	10	Ilyusha class minesweeping boats. 70 tons, 12 kts. Radio-controlled minesweeping drones.
1954-59	45	K-8 class minesweeping boats. 26 tons, 12 kts. Wooden hulls. Many built in Poland. Being replaced by Yevgenya class.
1983	–	"Baltica"—converted small vessel. Used as minesweeper auxiliary.

*Indicates years entered into service.

1984 – Possible new 40-ft. remote-controlled unmanned minesweepers.

Soviet aircraft and helicopters are used in a similar manner to those of the U.S. Navy. (See Figures 2.80 through 2.82 for fixed-wing aircraft such as the "May" ASW aircraft, and the Blinder A supersonic bomber that can be used in the minelaying roles. Also see Figure 2.83 and 2.84 for two Soviet helicopters, the Hip-8 and Mi-14 that have been utilized by the Soviet Navy in mine-clearing operations, primarily in the Middle East.)

- MI-8 "Hip" helicopters used for mine countermeasures. In 1974 operated from A/S cruiser *Leningrad* in mine-clearing operations at the south end of the Suez Canal.
- MI-14 "Haze" helicopters are configured for mine countermeasures.
- Soviet fixed-wing aircraft that can lay mines:
  - Badger A & G
  - Bison
  - Bear F
  - Blinder A
  - Mail (Modified)
  - May

Vietnam

8 Ex-U.S.A. MSB-5 class minesweeping boats. 30 tons, 12 kts. Wooden hulls. Two sweep current generators. Can sweep magnetic, contact and acoustic mines.

8 Ex-U.S.A. MSR (Minesweeper, Riverine) class. 30 tons, 14 kts. Some were equipped with a pipe frame projecting ahead of the boat to explode contact mines.

8 Ex-U.S.A. MSM (Minesweeping Monitor) river minesweepers. 70 tons, 8 kts. Could sweep mechanical and acoustic mines.

10 Ex-U.S.A. MLMS (Motor Launch Minesweeper) class. 25 tons, 8 kts. Wooden hulls.

Yugoslavia

4 French Sirius class coastal minesweepers. 400 tons, 15 kts. Wooden plank hulls on metal framing.

4 British "Ham" class minesweepers. 123 tons, 14 kts. Wooden plank hulls on metal framing.

6 117 class inshore minesweepers. 120 tons, 12 kts.

3 Nestin class river minesweepers. 65 tons, 15 kts. Hulls of light metal alloy. Used on Danube.

14 Small river minesweepers. 38 tons, 12 kts. Used on Danube.

Australian Navy "Bay" class minehunter catamaran *HMAS Rushcutter* (M80). Inshore minehunter with good stability, and wide deck area.

Model of RAN catamaran minehunter

Figure 2.1

Figure 2.2. Belgium Navy's first Tripartite minehunter *Aster*, on commissioning at Ostend in December 1985.

Figure 2.3. Brazilian minesweeper, German built, Schutze class. Builder: Abeking & Rasmussen.

Figure 2.4. Finnish Navy minesweeper Kuha-22. Glass-reinforced plastic hull. Six built 1974-1975.

French Navy Minehunters

Class *Hull*	Circé *Laminated Wood Plastic*	Dompaire *Wood*	Eridan *Glass-reinforced Plastic*
Displacement (tons)	460	700	500
Length, overall (m)	50.9	59.3	47.1
Beam at deck (m)	8.9	10.7	8.9
Main Engine	1 Diesel × 1370 Kw	2 Diesel × 590 Kw	1 Diesel × 1400 Kw
Minehunting propulsion	2 active rudders	Main + 1 bow thruster	2 active rudders + 1 bow thruster
Max. speed, en route (knots)	15	13.5	15
Speed, minehunting (knots)	0-6	0-6	0-6
Range (mm)/at speed (knots)	3000/12	3000/10	3000/12
Complement (men)	47	54	49
Minehunting system	Skubermor I	Skubermor II	Skubermor III
Minehunting navigation	Toran/Syledis	Toran/Syledis	Toran/Syledis/Hifix
Minehunting piloting	manual	manual	automatic
Mine disposal	2 PAP-104	2 PAP-104	2 PAP-104
Max. operating depth (m)	80	80	80
Sweep	—	double Oropesa type	single Oropesa type

TABLE 1.

Figure 2.5. French Tripartite minehunter, the "Eridan" class. Built under supervision of French Direction des Constructions Navales (DCN). Belgium and Netherlands are other members of the Tripartite. 47.10 meters length, 8.90 meters beam, 2.49 meters depth. Mean displacement of 562 tons. GRP Hull. 15 kts. cruise speed. Sweeps at 8 Kts. Thomson CSF DUBM-21B hull sonar, EVEC 20 data handling system. PAP-104 submersible. Armed with 20 mm. gun.

Figure 2.6. Three views (see next page also) of the French Navy Tripartite class minehunters. The center view M642 is the *Cassiopee*. Belgium class is the *Aster* and Netherlands class is the *Alkmaar*.

Figure 2.6. *(cont.)*

Figure 2.7. French minehunter Tripartite class CMT *Cassiopee* during shock tests.

Figure 2.8. French Circe class minehunter. Built exclusively as a minehunter.

Figure 2.9. Model of the Batiments Antimines Oceaniques (BAMO). Designed for rough weather. Will extend French MCM capabilities to edge of continental shelf.

Figure 2.10. West German Navy minesweeper "Sirius," one of 21 305-ton "Schutze" class coastal minesweepers.

Figure 2.11. West German Navy "Lindau" class minehunter shown is the CL-331B "Tubingen."

Figure 2.12. West German minesweeper CL-341.

Figure 2.13. West German minehunter CL-331B.

Figure 2.14. West German Minehunter/Minelayer/Minesweeper CL-343 MCM Vessel Type MW-V 50. (See next page also.) Hull is constructed of non-magnetic steel-stainless. Three yards are involved in construction: Abeking & Rasmussen at Lemwerder near Bremen; Lurssen at Bremen-Vegasack; and Krogerwerft at Kiel. Each yard holds jigs for a one-third section of the hull and distributes sections to others for assembly. German yards have the skill for welding stainless steel without distortion. The vessels' acoustic and magnetic signatures, as well as resistance to shocks, exceed NATO specifications. By using no smoke stacks, a better IR and lower radar signature are obtained, plus view from bridge is improved.

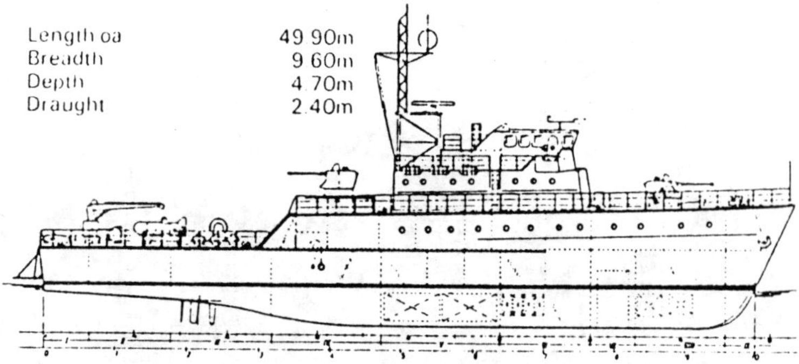

Figure 2.14 *(cont.)*

Figure 2.15. West German Navy F-1 "Troika" Drone. Bldr. is MAK at Kiel.

Figure 2.16. Manned Guidance Ship leading two of its remote-control "Troika" craft.

Figure 2.17. Two small remote-controlled "Troika" vessels lead the way through a minefield, simulating a large ship's magnetic and acoustic signature to detonate influence mines. Then the manned guidance ship cuts mooring cables in a mechanical sweep. Reduces the high risk involved in disposing of modern mines.

Figure 2.18. HMS Nurton (M-1166), a British Ton class minehunter.

Figure 2.19. HMS Ledbury (M-30) British Hunt class minehunter.

Figure 2.20. British River class minehunter/minesweeper.

Figure 2.21. British Hunt class minehunter. This type was deployed in the Persian Gulf. Hull is of glass-reinforced plastic.

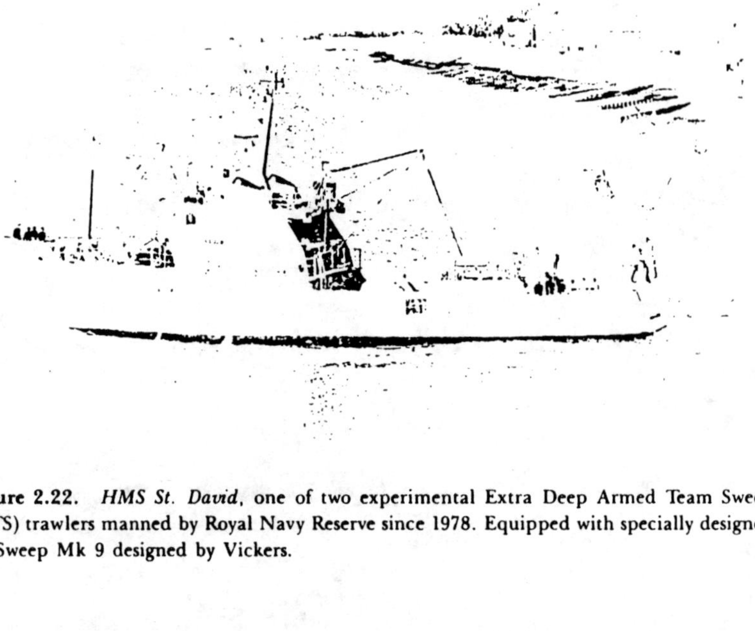

Figure 2.22. *HMS St. David*, one of two experimental Extra Deep Armed Team Swe((EDATS) trawlers manned by Royal Navy Reserve since 1978. Equipped with specially design(Wire Sweep Mk 9 designed by Vickers.

Figure 2.23. Two views of British Royal Navy 52-meter Vosper Thornycroft single-role minehunter, Sandown class.

Figure 2.24. Frontal view of British Hunt class MHSC *HMS Ledbury*. Such vessels have GRP hulls and are equipped with French PAP-104 mine disposal vehicles. This type saw service in the South Atlantic—Falklands Campaign.

Figure 2.25. A British Hovercraft Corp. BH. 7 Mk. 2 experimental minehunter. Potentially useful for operations in coastal areas.

Figure 2.26. Royal Navy Hovercraft conducting a series of tests at HMS Osprey, Portland, Great Britain. Tests were designed to evaluate the Hovercraft's mine countermeasures advantages of invulnerability, high speed, high controllability, and lower cost. This craft has been shown to have low underwater signatures—magnetic, acoustic, and pressure—and therefore is unlikely to explode mines. The Hovercraft was virtually immune to underwater explosions. Was fitted with Plessey Marine 193 sonar, 2048 Speedscan installation, and Racal-Decca Trisponder navigation equipment.

Figure 2.27. The prototype Italian Navy Lerici lying at Intermarine's Sarzana shipyard during fitting-out.

Figure 2.28. Italian 760-ton MCMV with GRP monocoque construction.

Figure 2.29. The SAPRI (M5551), one of four Lerici 2a Series Minehunters/minesweepers that are now operational with the Italian Navy. Intermarine of the Montedison Group of Italy now has contract with Hercules of the U.S.A. to build for the U.S. Navy at the Sayler Marine facility in Savannah, Georgia.

Figure 2.30. Japanese Navy "Hatsushima" class coastal minesweeper. Armed with 20 mm. Oerlikon gun. Later vessels may have General Dynamics Phalanx CIWS.

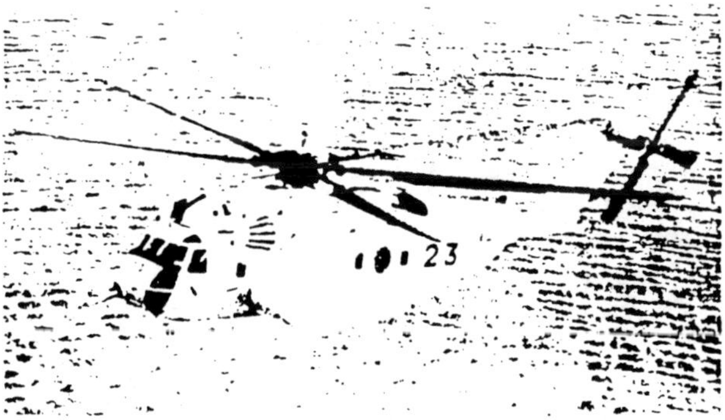

Figure 2.31. Japanese Maritime Self-Defense Forces S-80M-1 Airborne Mine Countermeasure helicopters. One of six delivered in 1989 from Sikorsky, who also built the U.S. Navy MH-53E. Japan desires a total of 12.

Figure 2.32. Various views (see next page also) of Royal Netherlands Navy Tripartite design Alkmaar class GRP minehunters. Netherlands provides the main propulsion system, single-shaft Werkspoor Diesels.

Figure 2.32 *(cont.)*

Figure 2.33. Norway announced in August 1987 that it would build ten surface-effect coastal minesweepers/minehunters, based on a civilian design. Construction is scheduled to begin in 1989, with delivery of the last vessel in 1996.

Figure 2.34. The *Koster*, third of the Swedish Navy's Landsort class GRP minehunters. Use Thomson-CSF TSM2022 minehunting sonar.

Figure 2.35. Swedish Navy Landsort MCMV uses Racal-SMS MAINS minehunting system as part of PEAB 9MJ 400 mine countermeasures system.

Figure 2.36. Swedish Landsort class *Kullen* in Goteborg.

Figure 2.37. Royal Swedish Navy "Alvsborg" class minelayer *Visborg*. Can carry a helicopter and is armed with three 40-mm. Bofors guns.

Figure 2.38. Swedish Coastal Minesweeper *Arko* (M57). commissioned in 1958, firing its Bofors 40-mm. L/70 gun.

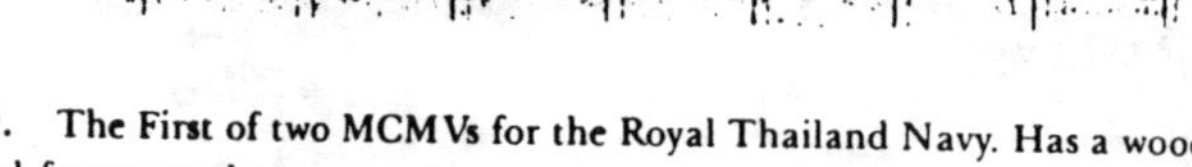

Figure 2.39. The First of two MCMVs for the Royal Thailand Navy. Has a wooden hull and amagnetic steel frames and superstructures.

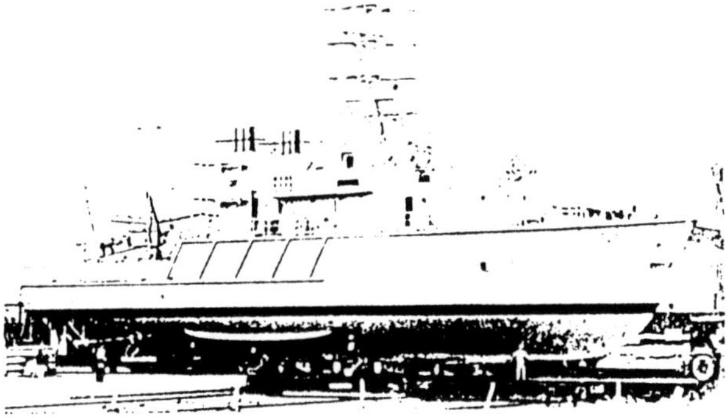

Figure 2.40. The Second of the Royal Thailand Navy's minehunters. During minehunting operations, the vessel moves at a speed of 7 kts. using its diesel electric auxiliary powerplant.

Figure 2.41. U.S. Navy minesweeper *USS Fearless* (MSO 442).

Figure 2.42. *USS Excel* (MSO 439). 172 ft long. Crew of 6 Officers and 65 Enlisted men. Five of the six U.S. Navy minesweepers that were on duty in Persian Gulf were Navy Reserve ships.

Figure 2.43. YP-676 class Yard Patrol craft. YP-677 had conversion for service with Craft-of-Opportunity Program at Peterson Builders, Inc. at Sturgeon Bay, WI.

Figure 2.44. U.S. Navy Coastal Minesweeper *USS Albatross* (MSC 289). 144 ft long. Crew of 4 Officers and 35 Enlisted men.

Figure 2.45. U.S. Navy Minesweeping Boat (MSB 47). 57 ft long. Crew of seven Enlisted men.

Figure 2.46. U.S. Navy Minesweeping Launch (MSL 28). 36 ft long. Crew of five Enlisted men.

Figure 2.47. U.S. Navy mine countermeasures ship now entering service. Both Peterson Builders, Inc. and Martin Marinette of Wisconsin are building these vessels. Avenger class.

Figure 2.48. Sideways launch for the *USS Sentry* (MCM-3) at Peterson Builders' Yard at Sturgeon Bay, Wisconsin.

Figure 2.49. Structural details of the Avenger (MSM-1) design showing wooden ribs and planking, with GRP coating.

Figure 2.50. Former U.S. Navy MSH-1 Cardinal class. Twin hull with rigid sidewalls, derived from SES designs. Cancelled by U.S. Navy Bell Aerospace Textron Design.

Figure 2.51. U.S. Navy MHC-51 Osprey class – modification of Italian Lerici class design vessel.

Figure 2.52. *SS Pratt Victory*, a WW II Liberty Ship used to run through Japanese waters after the war to confirm that pressure mines were still active. One of three of the "Guinea Pig Squadron" was controlled from on deck, with crew members standing on mattresses or raised wooden crates to lessen the explosive shock if the ship exploded a mine. December 1945.

Figure 2.53. Russian minelaying submarine "Krab," 1910.

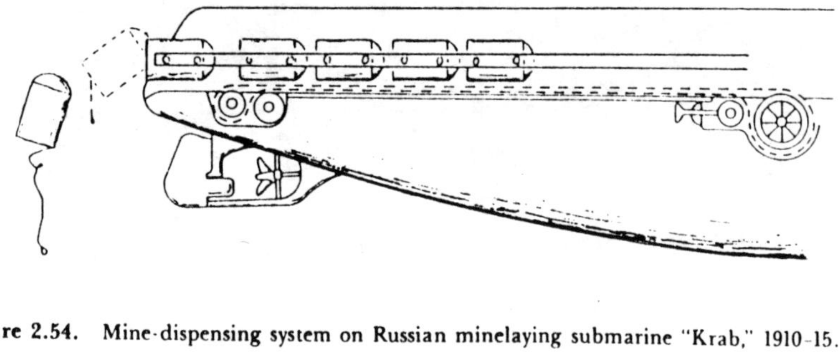

Figure 2.54. Mine-dispensing system on Russian minelaying submarine "Krab," 1910-15.

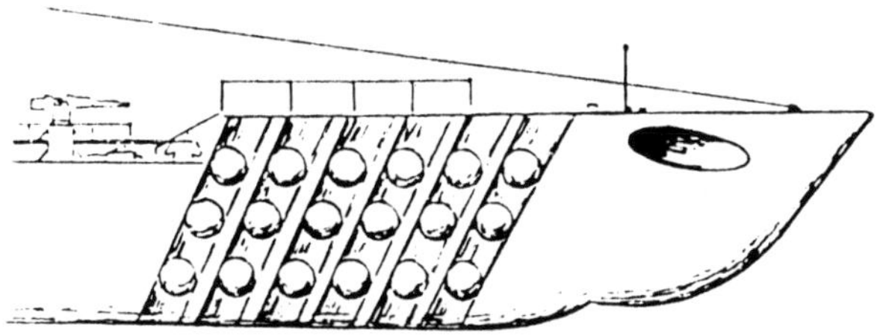

Figure 2.55. German UC submarine, 1915.

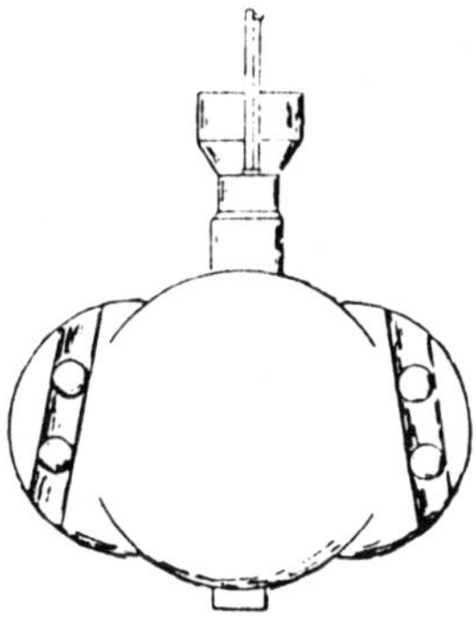

Figure 2.56. British E class submarine, 1915.

Figure 2.57. *USS Argonaut* in New York, March 1929.

Figure 2.58. *USS Argonaut*, U.S. Navy's largest submarine in Washington Navy Yard, November 1928.

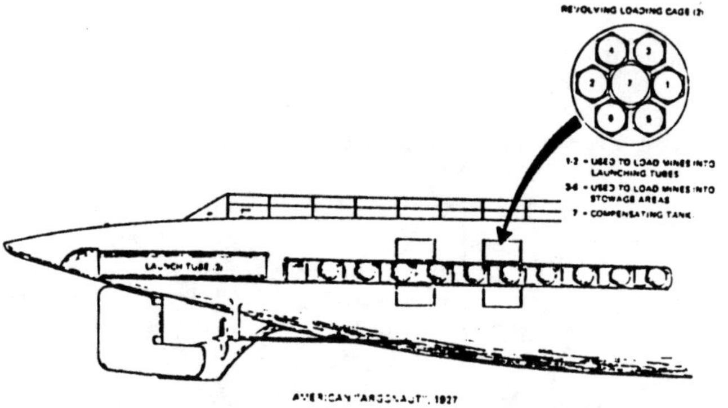

Figure 2.59. Mine-handling system on *USS Argonaut* included revolving cages to move mines from storage racks (3,4,5,6) into launching tubes (1,2).

Figure 2.60. A typical Soviet submarine can carry 50 mines.

Figure 2.61. Most Soviet submarines can lay mines. Above, Soviet Navy "Yankee" class SSBNs. As missile tubes are removed to comply with SALT treaty, these submarines could be modified to carry a large number of mines.

Figure 2.62. High-speed minelaying by Navy aircraft.

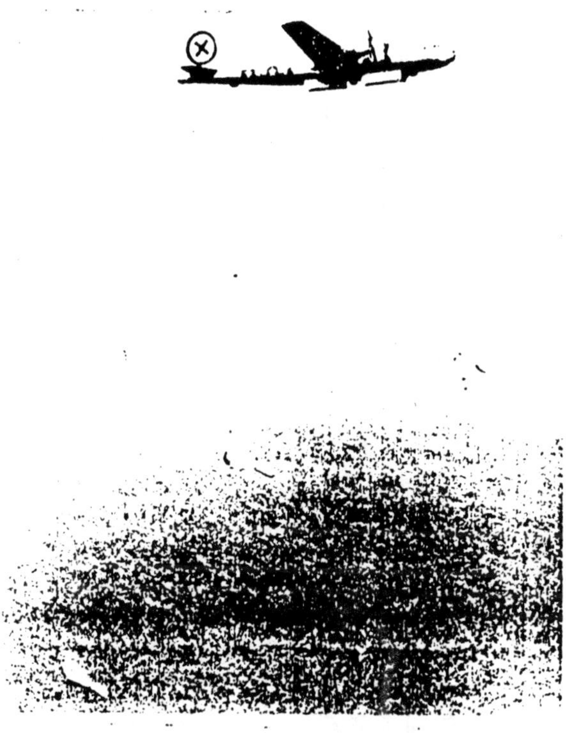

Figure 2.63. View showing World War II B-29 U.S. Army bomber aircraft dropping mines at Shimonoseki Straits in Japan.

Figure 2.64. U.S. Navy A-6 attack aircraft, which did much of the minelaying in North Vietnam. Numerous weapons-carrying stations permit this aircraft to carry a good load.

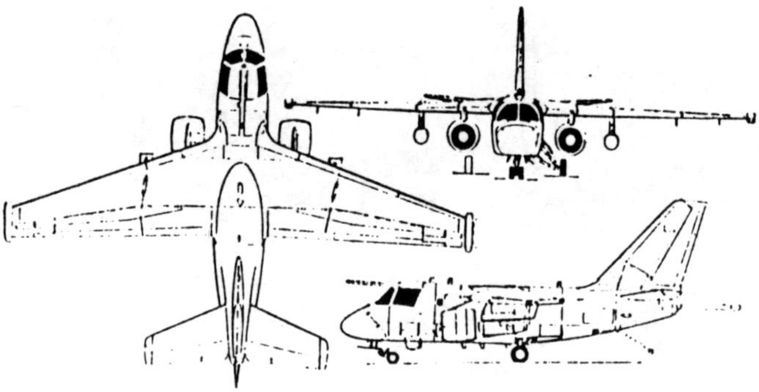

Figure 2.65. U.S. Navy S-3A ASW aircraft that can lay mines. This aircraft is normally operated from aircraft carriers.

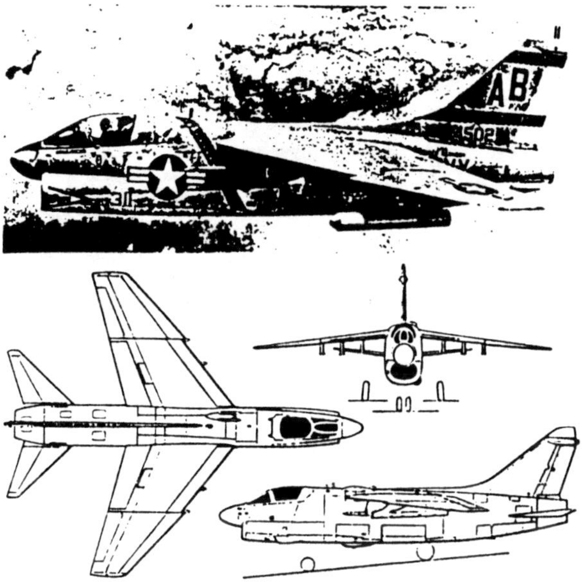

Figure 2.66. Views of the U.S. Navy A-7 attack aircraft. This aircraft can lay mines.

Figure 2.67. A U.S. Air Force C-130H Hercules cargo aircraft dropping a dummy defensive mine in recent test.

Figure 2.68. U.S. Navy Maritime Patrol ASW aircraft that can also lay mines.

Figure 2.69. U.S. Navy RH-53E Sea Stallion minesweeping helicopter clearing mines with EDO MK-105 sled in Suez Canal after the 1973 Middle East War.

Figure 2.70. U.S. Navy MH-53E helicopter pulling sled to clear mines. Eight MH-53Es have joined fleet squadrons since 1986.

Figure 2.71. Soviet "Sonya" class wooden hull minesweeper. Over 50 were built for the Soviet Navy.

Figure 2.72. Soviet "Vanya" class coastal minesweeper.

Figure 2.73. Soviet "Yevgenya" class inshore minesweeper. GRP hull. Carries a small dipping sonar aft.

Figure 2.74. Soviet T-58 class ocean minesweeper. 4-57 mm. twin AA guns. Built from 1957 to 1964.

Figure 2.75. Soviet T-43 class ocean minesweeper. 4-37 mm. twin AA; 4-25 mm. AA twin guns. Built 1948-1957. Many transferred to Eastern Bloc countries.

Figure 2.76. Soviet "Krivak" I class ASW Frigate. Stern view shows VDS housing, roller openings for mine rails, and towed torpedo decoys in racks.

Figure 2.77. Soviet "Natya" I minesweeper—Ocean Fleet minesweeper first reported in 1971. Planned as a successor to "Yurka" class.

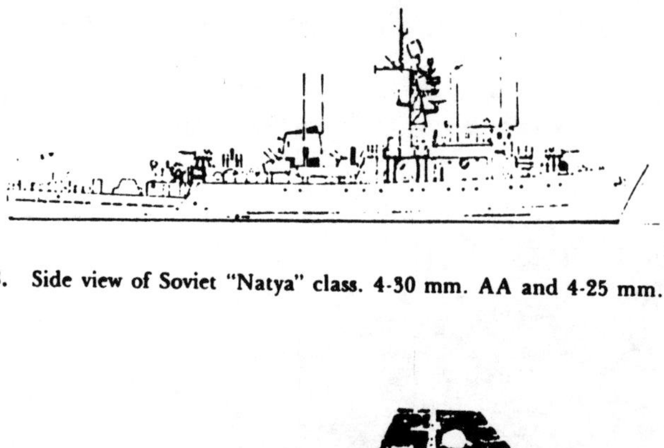

Figure 2.78. Side view of Soviet "Natya" class. 4-30 mm. AA and 4-25 mm. AA guns.

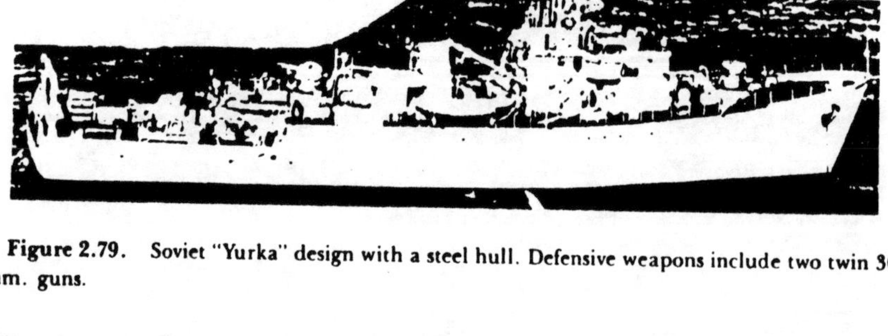

Figure 2.79. Soviet "Yurka" design with a steel hull. Defensive weapons include two twin 30 mm. guns.

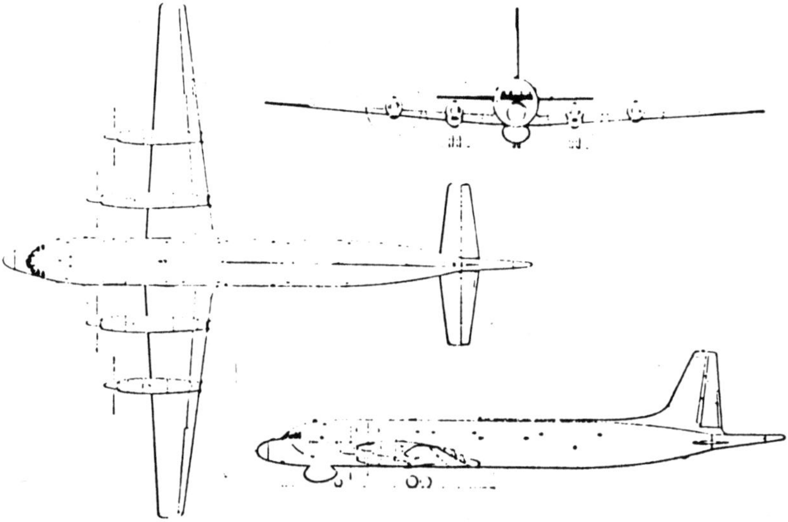

Figure 2.80. Soviet ASW maritime patrol aircraft that can lay mines. Ilyushin Il-38 is a derivative of the Il-18 airliner.

Figure 2.81. Soviet Il-38 maritime patrol aircraft "May." Four Ivchenko AI-20 turboprop engines.

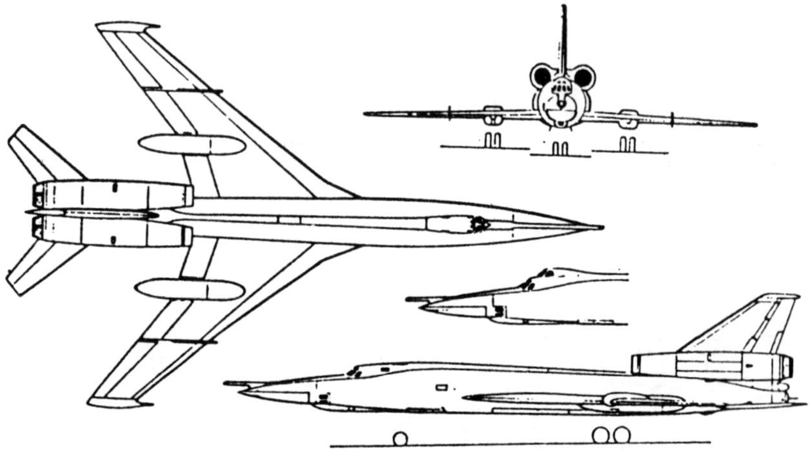

Figure 2.82. Soviet Tupelev TU-22 twin-jet supersonic bomber (Blinder A). Can deploy minefields.

Figure 2.83. Soviet "Hip-8" (Mil) transport helicopter. Can carry 4 tons internally. Soviet Navy flies a number of these in amphibious assault, minesweeping, and utility roles. Hip-C is the production version. This helicopter was observed flying from the helicopter carrier *Leningrad* in 1974 in mine-clearing operations at the southern end of the Suez Canal. Also flown by Warsaw Pact Military Services. Clamshell doors at rear of cabin. Two turboshaft engines. Five-blade main rotor.

Figure 2.84. Soviet Navy Mi-14 (Haze) Helicopter has an MCM as well as ASW role.

CHAPTER 3

Naval Mines of Various Nations

GENERAL DESCRIPTION—MINE CHARACTERISTICS

Mines, mine mechanisms and mining mentioned herein refer to sea mines. That is, those mines which are placed in a marine environment, whether in deep water, shallow coastal areas, harbor entrances, or a riverine location of rivers, canals and estuaries. "Destructors" is a class of mines that are essentially general-purpose bombs having mine mechanisms as their fuzing devices and that can be used as land mines as well as sea mines. Mines are designed for use in many situations, against different target classes or types of ships. To adapt the mines to a wide variety of missions, they are becoming more complex. Many different kinds of mines and variations thereof exist in the nations' stockpiles.

Some mines, with small explosive charges, are designed primarily for use against river boats and wooden vessels of small displacement. Other mines will destroy or damage most major combatants. Some mines were designed for use against submarines. A complex mine is complex mainly because of the "intelligence" that is built into its firing system.

Mines can be described as being defensive or offensive. When used defensively, mines may be placed in our own and friendly ports, harbors, anchorages (perimeter defense), channels, bays, estuaries and open water, to protect against enemy offensive seaborne attacks into these areas. The threat of such mines is often of equal importance with the actual sinking of ships, since their very presence or threat requires countermeasures to sweep or neutralize them, causes delays in shipping schedules, and can cause shipping to be detoured over longer routes that could be more dangerous due to submarines. When used offensively, mines are placed in the enemy's waterways, anchorages, harbors, and channels or may be planted in sea lanes removed from the enemy's harbor area focal points to attack his military and commercial shipping.

Most mines are classified as the independent type. Once placed and armed, they can be actuated by the presence of a ship, without any human control. Controlled mines have an electrical system connecting them to a shore facility. Controlled mines are not used at present by most navies.

Mines are also classified according to the position that they assume in the water:

- Bottom Mines (also known as Ground Mines)
- Moored Mines
- Drifting Mines

Note: Drifting mines were limited by the Hague Convention of 1907 and are not in the U.S. Navy inventory. However, nations such as Iran have recently used them in the Gulf of Suez, Persian Gulf area.

Bottom mines are used most effectively in relatively shallow waters. A large negative buoyancy or tendency to sink brings the bottom mine to rest on the ocean floor and keeps it there. In very deep water, surface vessels may pass over the mine without actuating its firing mechanisms or, in the event of an actuation, without suffering much damage. The bottom mine planted in deep water can still be very effective against submarines.

The explosive charge and firing mechanism in a moored mine are housed in a positive buoyancy case, or one that tends to float. A cable or chain attached to a negative-buoyancy anchor on the sea bottom holds the case at a predetermined depth below the surface.

The drifting mine floats freely at or near the surface. It has no anchoring device, and has an approximate neutral buoyancy. This is the type recently used by Iran.

For classification by the method delivered, mines fall into three groups:

Surface-laid
Submarine-laid
Aircraft laid.

With appropriate modifications, aircraft-laid mines with no flight gear, and submarine-laid mines can be placed by surface craft, and surface-craft-laid mines can in some cases be placed by helicopters.

As was done in North Vietnam, aircraft-laid mines are normally employed in offensive operations and are dropped from aircraft in the same may as a bomb. They can be hung on aircraft ordnance wing stations or any station with an appropriate lug for attachment and release. Aircraft provide the capability for reseeding minefields over an extended period of time without danger from going into a field that has been previously laid, such as would be the case with surface ships or submarines. Aircraft can also be used for mining enemy-held inland waterways.

Submarine-laid mines are normally used in offensive operations. They are specially configured so that they can be launched from submarines. The submarine can carry only a limited number of mines unless it has been configured primarily as a minelayer. Normally, torpedoes and missiles take up much of the submarine storage space. Minelaying operations by the submarine can be very covert and the mines can be placed at enemy ports or operating areas at great distances from friendly bases.

Surface laying of mines is the most economical method of delivery because of the greater number of mines that can be carried each trip in the vessel. Most air- and submarine-laid mines can be adapted for surface laying if need be. If the enemy controls the sea area in which the mine field is to be laid, obviously the surface laying poses the greatest risk. Either offensive or defensive mines can be placed by surface vessels, depending upon their employment.

All modern mines are activated by "influence." Influence mines do not require physical contact with the target. They are activated to fire by the presence of the target as sensed by magnetic, acoustic, or pressure devices.

Mines can be further classified according to type:

Service
Nonservice
Test

A service mine is an explosive weapon used against wartime targets. It consists of an explosive-loaded case, explosive initiators, mine mechanisms, and auxiliary components.

A nonservice mine is a reusable configuration primarily used for training exercises. Empty cases, or inert-loaded cases are used, but small explosive devices and perhaps pyrotechnics can be utilized that enhance realism in training.

Upon detonation of a mine by an actuation device, the charge is converted into a hot mass of gas at high pressure. This expands and generates a spherical, sharp-fronted shock wave. The gas bubble formed then collapses to generate a second shock wave. The two waves ultimately vent at the surface, the peak pressure decaying inversely with the distance from the explosion.

When the pressure wave reaches the target's hull, the dynamic effect can actually be doubled as the shock wave is reflected and transmitted along the hull. With moored mines, the bubble will frequently vent before it is fully developed, but ground mines benefit from sea bottom reflection. Nuclear charges are especially effective with a typical 1 KT charge generating an overpressure of 300 kg/cm^2 at 0.5 km, 50 kg/cm^2 at 1 km and 20 kg/cm^2 at 2 km. It should be remembered that the pressure pulse of any mine is modified by the "surface cut-off" when the positive shock wave is reflected at the point where the air and sea meet, creating a negative pressure wave that causes pressure at points below the surface to be cut off suddenly, with the result that shallow-draft vessels are less liable to damage than deep-draft ones.

MAJOR COMPONENTS

This pertains to the U.S. Navy configuration of mines, but it is representative of what would be found in the mine inventory of other countries. The primary components are:

Case
Explosive filler
Anchor (if used)
Firing mechanism
Accessories

Note: See Fig. 3.1 for a view of a typical mine

Most of the standard naval mine cases can accommodate a choice of the many firing mechanisms and accessories. Therefore, the "mark" designation of the mine usually applies only to the general configuration of the case or case and anchor, while Mod, a specific modification of a mine, normally designates its type of firing influence mechanism. A minor difference, such as the flight apparatus used, would be denoted by an operational assembly (OA) number for the particular MK-Mod mine. Thus, a mine might be typically identified as follows: Mine Mk 52 Mod 6, OA04.

Accessories, or auxiliary components ensure the maximum operational effectiveness of the mines. An example of one of these is the mine safety device. They are used to protect personnel who handle mines prior to and during minelaying operations. They also protect the minefields themselves so that movements of the mines in their planted environments immediately after placement do not cause spurious detonations. Accessories also include batteries which furnish the power for the operation of the mine; an arming device which could be a hydrostatically operated safety device that withholds the detonator from the explosive train until the mine has reached a predetermined period, permitting safe infiltration by a friendly operator; a ship counter which delays firing until a preset number of complete cycles of the firing mechanism has occurred, thus giving a sense of false security to vessels passing near the mine; and, in the case of aircraft-laid mines, flight gear which decreases the impact velocity of the mine as it strikes the water. Flight gear accessories include parachutes, release mechanisms, control units, nose fairings and tail fins. There are also various types of automatically operated anchors for moored mines, but the anchors are not considered to be accessories. The flight gear accessories are external to the mine itself and are used to reduce air drag and buffeting of a mine carried on an external aircraft station, and to reduce the impact velocity of the mine.

Aircraft, ships, submarines and all types of U.S. Naval activities and commands whose missions include mining responsibilities take part in recurring tests. These Fleet Service Mine Test Programs are designed to provide statistics on mine reliability and readiness. Fleet Service Mine Test (FSMT) mines for these tests are actual service-mine configurations in which the main charge is inert rather than explosive, and test conditions approximate actual service environments as closely as possible. Nonservice mines, unlike FSMT mines, can have their actuating circuits modified to operate, for test purposes, devices which are not part of the service mine configuration. The nonservice mines include smoke signals released later to locate the mine and provide recovery via a mooring line, obviating any need for the use of divers. New mine-actuation simulators have been designed for use with the newly designed mines.

ACTUATION METHODS

A mine is actuated when firing mechanisms receive a small impulse from the search coil, magnetometer, hydrophone, or pressure device. The firing mechanism must process the signal as a valid target.

It is only natural that newer and better firing mechanisms will be designed and used to avoid actuation by enemy countermeasures. The original signal still must meet the preset criteria to trigger the firing circuit.

An influence mine is actuated by the proximity of a ship under the proper conditions. The three classes of influence firing mechanisms are:

Magnetic
Acoustic
Pressure

Magnetic

Every steel ship has a definite magnetic field about its hull. The existence of this field can be used to actuate a mine-firing mechanisms sensitive to it. In World War I, some vessels were built in the United States with concrete hulls. These were supposed to be able to operate in fields of magnetic-type mines. The Federal Republic of Germany has built and is continuing to build minesweepers using a steel that is largely anti-magnetic. This is, again, a distinct advantage when operating around magnetic mines.

The induction-type mechanism employs a search coil and a means of amplifying the signal from the search coil. Some induction mechanisms use a sensitive relay for this purpose. Others use more sophisticated electronic circuitry to process the input signal. Induction firing mechanisms can be used in mines with ferromagnetic cases. Magnetometer mechanisms in moored mines require mine cases of nonmagnetic material.

The magnetometer magnetic firing mechanisms measure changes in the earth's magnetic field caused by the proximity of ferrous material. The changes may be either single or three axis. Magnetometer inputs are used in the newer magnetic mechanisms in conjunction with advanced signal processing techniques. See Fig. 3.2 for typical magnetic signature.

Acoustic

Sounds such as propeller noises, machinery noises and hull vibration usually accompany the passage of a vessel through the water. Total disturbance depends on such factors as ship size, type and number of propellers, type and speed of engines, condition of loading, character of the bottom and depth of water. A vessel's acoustic signal is then variable according to a complex number of operating and physical conditions. Acoustic mines must be designed to prevent an intense signal from actuating the firing mechanism at distances beyond the effective radius of the explosive charge. They must not be responsive to certain types of underwater sound not produced by targets, such as a school of fish. Mines must also not be easily actuated by countermining noises.

Acoustic mines are usually fired by mechanisms that actuate only when the sound level reaches a certain value calculated from building up at a prescribed rate of change. If the incoming sound builds up too quickly, as in the case of an underwater explosion, an anticountermining device would inhibit firing. If the sound builds up too slowly, the mine will not respond at all. See Fig. 3.3 for typical acoustic signature.

Pressure

As a ship moves through the water there is a continuous flow of water from the bow to the stern of the moving ship. As the forward part of the moving ship displaces water, an equal amount must flow to the after part of the ship to restore the displaced water after the ship has passed. This continuous water flow creates variations in the pressures at various depths. The pressure differential becomes more pronounced when the ship is moving in confined waters such as rivers, but is still appreciable in the open sea even at considerable depth. The pressure signature of a ship is a function of its speed, displacement and the water depth. See Fig. 3.4 for view of typical pressure signature.

Ocean swells and surface waves also produce pressure variations or "signatures" of considerable magnitude. To avoid premature firing because of wave action, pressure-firing mechanisms are designed to ignore rapid pressure fluctuations. Pressure-firing mechanisms are seldom used alone, but are generally combined with other influence-firing devices. Some mines can contain combinations of influence devices. This combination consists of any two of the three influences, or all three.

DEPLOYMENT AND OPERATION

During World War I, surface ships and submarines were used to plant mines. Some had thought during that war to use aircraft to plant mines, but it was not until World War II that mines were successfully planted in quantity by aircraft.

The advantages of air delivery are that aircraft can lay mines rapidly and in large quantities. Aircraft are the only laying agents capable of replenishing a large field without danger from the field itself. They can easily lay mines in shallow bodies of water, including rivers and harbors, which cannot be penetrated by submarines or surface minelayers. Good examples of aircraft mining were the mine blockades of Japan in WW II and Haiphong Harbor in North Vietnam.

Most air-laid mines use some sort of flight gear to decrease their water impact velocity. This usually consists of a parachute, parachute pack, and release equipment. As the mine strikes the water, the release gear frees the mine case from the parachute. The parachute and mine then sink free of one another. A nose fairing may be used on an air-laid mine to reduce the drag during carriage.

Almost any aircraft that can carry bombs can also lay mines. This can include Navy, Air Force, Marines, Army, Coast Guard, etc. Like bombs, air-laid mines are assembled with arming wires threaded through delay devices which keep the mines unarmed until they have reached certain water depths. To release the mine for arming, the pilot or bombardier operates controls which cause the wires to pull free of the mine at the instant of release from the bomb racks. To jettison mines safely, he lets the wires fall with them. In addition to their parachutes, most air-laid mines have fins, and some have fairings. These improve streamlining and thus make the trajectories easier to predict.

The only effective way to lay large minefields in enemy-controlled waters is by aircraft, which can sow large areas in a short time. In 1987, a B52D Air Force bomber demonstrated its minelaying capabilities to a European NATO audience. The S-3 Viking ASW aircraft, P3 Orion ASW aircraft, the A-6 Intruder, and A-7 Corsair attack aircraft have a similar wartime capability in the U.S. Navy.

Other nations of the world that have a Navy or Air Force with suitable aircraft can also lay minefields.

The submarine is the best mine-laying vehicle if secrecy is of importance. Submarines can carry mines great distances from a home base. If the mines are self-propelled by torpedo-like vehicles, the submarine can launch them into enemy-held harbors and inland waters without risking a complete penetration of an area for mining.

The submarines of most countries can lay mines through their torpedo tubes, and many manufacturers produce such cylindrical 533mm diameter mines. A notable exception is the West German series of small submarines designed by IKL Co. of Lubeck,

West Germany. As their swim-out tubes cannot launch mines, IKL has designed an external "necklace" dispenser. Although submarines can and do lay mines in hostile water, the numbers can never be sufficient to pose a major threat.

At the outbreak of WW II, the torpedo was considered the primary weapon of the submarine in the U.S. Navy on missions in the Pacific against Japan. However, the results of specific tactical mining missions by the Seventh Fleet submarines changed the loading policy to include a combination of mines and torpedoes. Submarines planted 576 mines, resulting in 27 ships sunk and 27 damaged, or one ship per 10 planted mines. This seems to indicate that when properly used tactically, the submarine-laid mines are effective and economically sound weapons.

During World War II, surface craft were used primarily for defensive operations, where mining is for protection of waters from penetration by enemy vessels. Planting was usually done from specially designed minelayers or from certain other surface craft. By working as a team, surface vessels can lay a large field in a short time. Thousands of these mines were laid during WW II. Fields were planted just outside Chesapeake Bay, near Cape Hatteras, and around Key West to protect our shipping against enemy submarines. Also, large fields were laid off North Africa and near Sicily in the Mediterranean.

Although there is no record of any of the enemy's ships being sunk or damaged in the U.S. Navy's defensive fields, neither is there any record of enemy ships or submarines passing through the fields. Possibly the knowledge that the fields existed prevented any attempts being made, proving the adage that mines work well even when they don't work at all. There have been cases where friendly ship captains who were not properly briefed or aware of friendly U.S. minefields went into such fields and were sunk. The burden here is to keep friendly ships advised of the proper channels for safe navigation.

Clandestine operations such as that mounted by a Libyan ro-ro ferry against the Egyptians in the Suez in 1984 show that surface minelaying is still feasible but, in wartime, large-scale surface minelaying as an offensive measure would be almost impossible. However, defensive fields would rely almost exclusively on such ships. Broad vehicle-decks are ideal for laying large numbers of mines.

Apart from the actual laying of mines, it is a comparatively simple matter to completely disrupt an enemy's seaborne trade by announcing that minefields have been laid, which in fact may not have been done. The enemy will be put to a great deal of trouble and expense before he discovers the ruse, and he will have to employ a large number of Mine Countermeasures Vessels (MCMV) that might be required to deal with an actual minefield. In a false minefield, it is only necessary to lay a very few mines and to explode an occasional one, by a time fuse, to keep the enemy guessing.

A SUMMARY OF VARIOUS NATIONS WITH A MINE WARFARE CAPABILITY, LISTING MINE TYPES

Note: This list is not all-inclusive but is meant to demonstrate most of the countries with capabilities. In some cases, older mines—even those no longer in inventory—are listed. This is for historical purposes and to enable the reader to follow the evolution of mines in that particular country.

Argentina

Locally made copies of a German Second World War type were found in Port Stanley harbor in the Falklands (Malvinas) by Royal Navy mine countermeasures ships after the Argentine surrender in June 1982. These were horned mines, operated by electrochemical or simple electrical switches protruding from a buoyant casing. The Argentine Navy also had available to them some British and other influence mines.

Australia

Royal Australian Navy has ordered from Marconi Underwater Systems Ltd. of Britain a number of "Stonefish" naval exercise mines.

Stonefish is designed for training of naval personnel in the detection and handling of mines. It can be programmed to imitate the characteristics of all known potential enemy mines. Incorporates latest microprocessor technology. The system incorporates a vessel data-gathering capability in its multi-mode programmable electronics which is capable of recording up to 50 individual exercises in real time during a six month in-water life. While still operational, and on the sea bed, remote adjustment of algorithms and threshold settings can be made to enable this exercise mine to emulate other mine types. After initiation of the recovery sequence, the mine can be recovered and readied for use within two hours. The life of the Stonefish is claimed to be in excess of 20 years.

British *(See Figures 3.5–3.11)*

During World War I, the British and U.S.A. planted 71,117 mines in Great Northern Mine Barrage. These were antisubmarine antenna mines (with copper antennas). On sweeping, only 21,295 were detected. About 20 to 30 kg of copper was used for the copper antennas.

MK-XX — Antisubmarine copper antenna mines.

MK-XXII — Antisubmarine copper antenna mines. Royal Navy is currently upgrading these by providing modern explosive and an actuation system based on BAe's (British Aerospace) Sea Urchin ground mine.

The British used an induction type firing mechanism, while the Germans used a magnetic firing mechanism for moored mines. These were actuated not only by a vessel, but also by the sea and currents.

It is doubted that the British produce any moored mines at present.

NATO gave the British the responsibility for developing a new Intermediate Water Depth Mine (IWDM). The Bracknell Division of British Aerospace (formerly Sperry Ltd.) and Plessey Marine are working together on the project.

Stonefish — Made by Royal Ordnance Factory. Production by Marconi Ltd.

Limpet — An exercise mine. Made by British Aerospace Ltd.

Sea Urchin — Made by BAe Ltd. A small multi-influence ground mine in three different sizes that can be laid by ships, submarines or aircraft. Can have a 250 kg, 500 kg or 750 kg warhead. Can operate to 90 meters.

Crusader — Made by BAe and Plessey Ltd. working with the USA firm of Honeywell. Called the "Advanced Sea Mine." Previously called CSM (Continental Shelf Mine) in cooperation with U.S. Navy NGM (New Generation Mine).

Dragonfish	Made by Marconi Ltd. An anti-invasion ground mine. Its magnetic actuator is sensitive enough to detect the passage of an air-cushion craft overhead.
Stonefish	Made by Marconi Ltd. A large multi-influence ground mine. Can have weight in range of 205 to 900 kg depending on configuration.
Hammerhead	Marconi Ltd. new mine under development, influence-type ground mine.
VEM	Versatile Exercise Mine. Also made by Thomson-CSF of France, now also by British Aerospace. Sophisticated training device which records the degree of success of minehunting techniques, and provides realistic practice in laying.
BHM	Buried Hunter Mine. Made by Lawborough Consultants. A torpedo mine such as Captor or IWDM, but it is fitted inside a specially designed self-burying casing. The mine will dig itself into the sand, mud or silt until hidden. This can be done either immediately after laying or on command from a ship, aircraft or shore station. It is little larger than a conventional ground mine, and contains an electronics command package, a pumping system to remove sand, mud or silt from underneath, and the encapsulated torpedo with its own sensor and command package. It can be activated instantly or can remain inert for a predetermined period. It can also be used, if suitably modified, to record and transmit the sound signatures of passing ships, functioning as a simple surveillance system until it needs to be activated.

The British are working with the U.S. Navy to develop a new mine. It is mobile, deepwater capable of detecting and defeating submarines in the 1990s. U.K. system is called Continental Shelf Mine (CSM). This is an Advanced Sea Mine (ASM). U.S. system is called New Generation Mine (NGM).

The Royal Navy wants a mine capable of deployment to Continental Shelf depths of 600 feet to provide a protective barrier around approaches to vital British bases.

The U.S. Navy wants a more offensive and deeper-water mine for use in areas close to the Soviet Union. Its mission is "to attack the sanctuaries." The new mine will be able to maneuver into position. Once in place, a computer-based sensor system will track targets at long range, discriminate between the signals it receives so that targets can be specified, and home in on the chosen victim. The mine can also be remotely controlled so that it can be active or inert.

In the U.K., the effort is led by the Admiralty Research Establishment (ARE). In the U.S., the effort is managed by the Naval Surface Weapons Center at White Oak, Maryland.

British Aerospace in the U.K. is working on a rocket propulsion system for the new mine. Plessey from the U.K. is working with British Aerospace. The U.S. partner is Honeywell.

France *(See Figure 3.12)*

TSM-2510	Made by Thomson-CSF. A submarine-deployable mine. Employs sensor detonation techniques. Is discharged into the sea via the submarine's torpedo tubes.

MCT-15 Moored influence mine built by Thomson-CSF.

TSM-3510 Ground mine which is laid by submarines from their torpedo tubes and is probably a multi-influence mine.

TSM-3530 Ground mine which is laid by surface ships, dropped overboard from rails, descends to the bottom, its fall being retarded by a sea parachute.

Germany

World War II *(See Figures 3.13–3.15)*

Pressure Influence or "Oyster" mine. Detonated by water pressure when a large ship passed overhead. This mine could be laid only in shallow water. Almost unsweepable.

German Navy used ships filled with concrete to act as mine bumpers or "sperrbrecher." Some high-speed craft were tried to actuate the pressure mines. British used giant trimaran hulls (HMS Cyrus and HMS Cybele). Pressure mines were best dealt with by clearance divers who located each mine and destroyed it with a small explosive charge.

German aircraft laid mines around the Suez canal in World War II. This closed down operations through the canal for many weeks. German aircraft also laid mines (about 500) during the Allied landings at Normandy, France. These sank some Allied vessels.

Figure 3.14 shows the main mine types used by the German Navy during World War II. Figure 3.15 demonstrates how effectively a few mines laid by a few German submarines in World War II along the Eastern U.S. and Canadian coasts closed ports to shipping.

Post World War II *(See Figures 3.16–3.17.4)*

Figures 3.16 and 3.17 illustrate that the East German and Polish sailors trained extensively in mine warfare.

FG-1 Ground Influence Mine. Made by Fraun Werke. Has a sensor which detects "physical fields generated by the hull of a vessel," which could mean that it is a pressure, magnetic or acoustic type of mine, or all three.

G-2 This is a recently developed modern ground influence mine. It is configured with magnetic, acoustic and pressure sensors, is modular in construction and utilizes the latest in microprocessors. This mine was developed under a joint German-Danish program. The present manufacturer chosen after competition is Krupp Atlas Elektronik (KAE).

Most Navy submarines lay mines through their torpedo-tubes, and many manufacturers produce such cylindrical 533 mm. diameter mines. An exception is the West German series of small submarines designed by IKL of Lubeck. As their swim-out tubes cannot launch mines, IKL has designed an external "necklace" dispenser.

Italy *(See Figures 3.18–3.23)*

MM-80 General purpose, multi-influence ground mine. Made by Misar Co.

MRP Ground mine. Made by Misar Co.

Manta Anti-landing craft mine. Made by Misar Co. Charges for countermining mines. Also made by Misar Co.

TAR-6 & TAR-16 Moored mines. Made by Whitehead Moto Fides. Sown together in a group. Their explosive charges are linked together by a line to create a vertical barrier between 23 and 26 meters deep. The detonation of one charge will set off the others in the group. The system is said to be effective down to 500 meters.

Italy, a prolific producer of mines, has three leading manufacturers: MISAR, Valtec and Whitehead Motofides.

MISAR produces the Manta mine, usable over a depth range from 2.5 to 100 meters. It features a programmable activation device which renders it live for any selectable period for up to 63 days. Manta's shape is such that it is retained in its sea bed position, even in fast currents or tidal races. The charge weight is 220 kg or 240 kb. MISAR also produces the MR-80 with charge weights of 400 kg to 1130 kg. This is an influence mine with a multi-sensor charge actuation system. The body of the MR-80 is formed from plastic material. It is lightweight and can be deployed from aircraft. Can be autonomous or shore controlled either via an electric cable or acoustic transmitter. MISAR has produced a family of fuses for use in their mines.

Valtec produces VS SM600, a multi-sensor weapon which incorporates extensive processing capability to provide ship signature-recognition capability for use against medium to large surface vessels and submarines. Its warhead contains 600 kg of TNT and it can be laid by surface ships and submarines or para-dropped by aircraft.

Whitehead Motofides' WP 900/1 is a multi-sensor triggered mine with arming-delay, ship-counting and auto-sterilization facilities. In a training version, this mine can be programmed to become positively buoyant to aid in post-exercise recovery.

Iran *(See Figure 3.24)*

MO8 Type Tethered contact mine. Possibly made in North Korea. These are Soviet-designed mines of World War II type.

Sweden *(See Figures 3.25–3.26)*

Royal Swedish Navy K-11 (also known as Bofors MMI-80) moored mine. Used in large numbers for coastal defense. Laid principally in controlled minefields. Can also act autonomously. It can be laid at preselected depths between 20 meters and 200 meters, and will anchor itself automatically. Has an 80 kg charge which is effective against small- and medium-sized surface ships and submarines.

Rockan mine. Made by Bofors Co. (See Figure 3.25). The Swedish Navy's K-12 harbor defense mine (better known as Bofors Rockan GM1 100) is unusual. When launched, it glides through the water up to 50 meters away from the launch point, ensuring a good spread when laid in numbers. Weight is 190 kg. It can be rapidly transported, and in an emergency can be laid in harbors from small vessels. The Rockan ground mine is an anti-invasion weapon deployable from either vessels or the shore. Its hull design causes Rockan to glide a horizontal distance of twice the depth of the water in which it is laid, this depth being from 5 to 100 meters. The mine's sensor system is insensitive to its own movement and the mine cannot be swept by equipment which creates a homogenous magnetic field. Bofors is incorporating this sensor system into its deep-water moored mine, the MMI-80.

U.S.A. *(See Figures 3.27–3.54)*

Note: In the past, mine development policy in the U.S. Navy was not guided by the [illegible] develop many unneeded mines. This was not necessarily a fault of the Laboratory. U.S. development of mines followed those types and designs of British and German use. Sixty-five mines were given Mark and Mod designations. Approximately 34 were released to service but only 7 were used in combat in World War II. Many older as well as new types are listed here.

Mk-2	A defensive bottom mine planted by ship. Used for harbor defense and controlled manually. Targets were ships or submarines. Now obsolete.
Mk-6	A defensive or offensive moored contact-type mine planted by ship. Used galvanic antenna. Now obsolete. Many used in the North Sea Mine Barrage in 1918, 56,000 U.S. Mk-6 mines were planted in 14 deployments. No longer in stockpile; obsolete in design. Cdr. S. Fullinwider, U.S.N., played an important part in the development of mine warfare during WW I and WW II. He retired in 1914 but was recalled in 1917 to head Desk N in the Bureau of Ordnance where he was in charge of design, development and procurement of the Mark VI (now Mk-6) mine which was the antenna-type moored mine used in the North Sea in World War I.
M-3 and M-5	Older firing mechanisms. Were copies of German magnetic-needle mechanisms. Target widths are too small and critical with depth. M-5 used in a moored mine is too susceptible to prematuring due to currents and waves.
M-4	Magnetic firing mechanism. Was too insensitive for use except for targets in very shallow water. Easily swept. Used in bomb-type Mk-13 mine. Obsolete from start. Had severe restrictions on release speed and altitude.
M-9 Mod 1	Magnetic mine firing mechanisms.
Mk-10 Mod 1	A submarine-laid contact mine. It was a useful mine for attrition fields.
M-11	Magnetic mine firing mechanisms.
A-3	Acoustic mine. One of first acoustic types. Used in WW II in Pacific. This type was susceptible to explosive sweeping. Operated on propeller noise in the acoustic region. Aircraft-laid. Mod 2 has some provision for target selection. Target width is too wide. Now obsolete.
A-5	A later acoustic mine. Operated on ship hull vibration in the subsonic region. Aircraft-laid. Too inflexible. Could be swept explosively. Now obsolete.
A-6	A pressure-magnetic mine type. Operated on the reduced water pressure under a ship resulting from its motion. Similar to a mine used by the Germans on the Normandy beachhead in World War II. This mine was considered to be unsweepable. Aircraft-laid. Can be avoided by a ship moving very slowly.
Mk-12	A mine specially modified for use in Pacific during World War II. Submarine-laid. Moored influence mine.
Mk-13	Unsweepable pressure and vibration mine. Aircraft-laid mine. Mod 5 was an acoustic version.
Mk-16	Surface-laid, moored contact mine.
Mk-18	Surface-laid, bottom influence mine.
Limpet	An offensive mine, surface-ship-laid, time delay, contact type. Used against surface ships or submarines.

Mk-25 Aircraft-dropped, magnetic induction bottom type mine. Firing mechanism is M-11, Mod 4. 2.13 meters long, 571.5 mm diameter, 900 kg, parachute-deployed. Warhead is 550 kg HBX-1 explosive. Arming delay from 1.5 hrs to 90 days. Sterilization period 3 to 365 days. Ship counts 1 to 10. Sensitivity is adjustable. Bottom depth 10-120 mtrs. Case depth 10-120 mtrs. With kit, could be converted to low-frequency acoustic or pressure mine. Mk-25 Mod 1 is vibration kit. Mod 2 is pressure kit. Nine-foot parachute used. B-29 Army bombers in World War II dropped many of these in the Pacific.

Mk-26 1000 lb aircraft dropped mine. Bottom influence mine. Many dropped from B-29 bombers during World War II near Japan. Used six-foot parachutes for deployment.

Mk-27 Submarine-laid mobile mine. Shaped like a torpedo. Used in the Pacific in World War II.

Mk-35 DST (Destructor) magnetic and seismic. Used against sampans, Osa and Komar boats. 200 kg and 400 kg.

Mk-36 Air-dropped bottom influence mine. 460 kg. Used six-foot parachutes for deployment. Dropped by B-29 bombers. Shaped like bombs.

Mk-40 & Mk-41 Similar to Mk-36 but with more explosive charge, aircraft-laid. Many of the Mk-41 type DST were used in Vietnam. The mechanism uses thin film magnetometer with 12 gamma fire level. Magnetic storms from sun can fire at the 12 gamma level. These perturbations set off 4,000 mines in one week in Haiphong Harbor in North Vietnam. Thereafter, U.S. Navy used seismic firing method.

Mk-52 Air-dropped bottom magnetic-influence mine. 540 kg, 2.26 meters long by 338 mm diameter. An offensive mine, used against surface ships and submarines. Mod 1 is acoustic firing method, Mod 2 is magnetic induction. Mod 3 is pressure, magnetic induction, Mod 5 is acoustic, magnetic induction and Mod 6 is pressure, acoustic with magnetic induction. Firing mechanisms Mk-20, Mod 1 for Mod 1, Mk-20 Mod 1 for Mod 2, Mk-22 Mod 1 and Mk-20 Mod 1 for Mod 3, Mk-21 Mod 0 and Mk-20 Mod 1 for Mod 5, Mk-22 Mod 1 Mk-21 Mod 0 and Mk-20 Mod 1 for Mod 6. Uses 595 lbs of HBX-1 explosive.

Mk-55 Air-dropped deep bottom influence mine for use against surface ships and submarines. An offensive mine. 963 kg with 500 kg warhead, 2.91 meters long by 592 mm diameter. Explosive is HBX-1.

Mk55 Mod 2 uses magnetic induction firing with Mk-20 Mod 1 firing mechanism.

Mk-55 Mod 3 uses pressure and magnetic induction with Mk-22 Mod 1 and Mk-20 Mod 1 firing mechanism.

Mk-55 Mod 5 uses acoustic and magnetic induction firing with Mk-21 Mod 0 and Mk-20 Mod 1 firing mechanism.

Mk-55 Mod 6 uses pressure, acoustic and magnetic induction firing with Mk-22 Mod 1, Mk-21 Mod 0 and Mk20 Mod 1 firing mechanism.

Mk-55 Mod 7 uses dual channel magnetic induction firing method with Mk-37 Mod 0 firing mechanism.

Sensitivities are adjustable. Ship counts can be set from 1 to 30.

Mk-56 Air-dropped moored total magnetic field mine. Offensive type for use against surface ships or submarines. 933 kg, 3.51 meters long by 592 mm diameter. The firing method is total field magnetic induction 2 channel type with Mk-26 Mod 0 firing mechanism. The explosive is 150 kg HBX-3. Five levels of sensitivity are available. Ship counts can be set for 1 to 7. Arming delay is 1 hr to 90 days. Non-magnetic stainless steel. Parachute release, air drop 400 kts from 400 ft or 500 kts at 100 ft.

Mk-57 Submarine-laid moored total magnetic field offensive mine used against surface ships or submarines. 935 kg, 3.07 meters long by 533 mm diameter (same as 21-inch torpedo tube). Has fiberglass case with steel anchor. The firing method is total field magnetic induction 2 channel type with Mk-26 Mod 0 firing mechanism. The explosive is HBX-3. Five levels of sensitivity are available. Ship counts can be set for 1 to 7. Arming delay is 1 hr to 90 days.

Mk-60 Captor (Encapsulated Torpedo). Submarine-laid, surface ship or aircraft-dropped. Uses Mk-46, Mod 4 acoustic homing torpedo payload. Primarily ASW in function. 908 kg, 3.66 meters long by 324 mm diameter. A moored active/passive acoustic mine for offensive use against surface ships or submarines. Usually deep-moored. Explosive used is HBX-3, 44.5 kg warhead. Has arming delay and sensitivity is adjustable. Bottom depth is from 200 to 2000 mtrs, case depth is 160 to 400 mtrs. Expected life is about 6 months. Minimum spacing is 1500 meters.

The Captor uses a vertically moored capsule containing a detection and control unit, safing and arming system, sterilizing and scuttling circuits, and electrical power sources. A reporting buoy (radio transmitter) is dry stored in the capsule. The anchor weight unit consists of mooring cable, cable reel, brake and weight. Captor can be delivered by A6, A7, P3, S3, B52 aircraft, SSN submarines and surface craft. It can be planted to 2000 mtrs with case mooring to 500 mtrs. The Captor is laid on the sea bed and carries a passive sonar that continuously monitors the surrounding ocean and when it gets a submarine detection, it reverts to active sonar to measure the range. The mine then launches a homing torpedo (Mk-44, or Mk-46-Mod 4 or later Mk-50 type) that attacks the target. About 2000 Captor mines in inventory. Principally to be deployed in choke points in the ocean through which enemy submarines must pass to reach the open sea.

This mine was developed by the Naval Surface Weapons Center at White Oak, Maryland as a primary antisubmarine weapon of the 1980s. The Captor detection and control system can detect and classify enemy submarines to a range of 1,000 meters. The Mk-46 torpedo is released when the enemy submarine ventures within the range of the torpedo's homing mechanism. Surface ships should not trigger the Captor's sensors.

As time went on, in the 1970s the mission of the Captor was broadened and an antiship capability was added; and may incorporate the Quickstrike mine mission as well. The Mk-50, new ALWT (Advanced Lightweight Torpedo) is scheduled to go into the mine to replace the Mk-46. This will give increased depth, speed and maneuverability, reduced echo return (the "Stealth" principle applied underwater), lower radiated noise, improved countermeasures and better homing characteristics.

Some 2000 of the Captors have been produced since the late 1970s. This mine is made by the Goodyear Company (now Loral Co.). The Bendix Co. Electrodynamics Div. and Collins Co. are also involved with components.

Mk-61 Actuation Mine Simulator. An exercise and training device. Used in Fleet mine exercises for training in influence-mine countermeasures.

Mk-62 225 kg mine, aircraft-dropped.

Mk-64 900 kg mine, aircraft-dropped. Modified bomb with Mod Kit Mk-75. Magnetic or seismic type mechanism. Also known as destructors.

Mk-65 900 kg mine, aircraft-dropped. This mine was developed, tested and approved for service use, but never produced in any quantity.

Mk-67 SLMM (Submarine Launched Mobile Mine). Converted Mk-37, Mod 0 torpedo. 754 kg, 4.09 meters long by 483 mm diameter. An offensive mine used against surface ships and submarines, bottom influence type. May be sown at some distance from the parent launching submarine or other craft. Propels itself into shallow water, which the submarine cannot reach covertly. Upon reaching its required position, it sinks to the bottom and turns itself into a mine.

IWD (Intermediate Water Depth). Formerly PRAM (Propelled Rocket Ascent Mine). Anti-surface and ASW mine for launch from submarines, surface ships and aircraft. A moored mine that could be laid in deep water out to edge of continental shelf. A medium-depth mine. Manufacturers were General Electric and McDonnell Douglas. Planned 900 kg warhead, operating to 200 meters. A 225 kg shaped charge warhead was planned for use against Soviet Alpha, Typhoon and Oscar submarines. This program was cancelled in 1982.

Destructors

Developed in the mid-'60s for use in Southeast Asia. The Destructor (DST) consists of a modification Kit Mk-75 and a general purpose bomb. With the Mk-75 DST modification kit installed, a 50 lb (227 kg) Mk-82 bomb becomes a DST Mk-36; the 1000 lb (454 kg) Mk-83 bomb becomes a DST Mk-40; and the 2000 lb (900 kg) Mk-84 bomb becomes a DST Mk-41. The DST is a simple weapon system to maintain and assemble and is a major advance in mine warfare.

Mod 3	Magnetic
Mod 4	Magnetic/Seismic
Mod 5	Magnetic/Seismic
Mod 6	Magnetic/Seismic
Mod 7	Magnetic/Seismic

DST-36 Quickstrike. A converted Mk-82 bomb as shown above. Quickstrike indicates rapidly assembled and placed into use. Aircraft-dropped. Magnetic bottom mine with 87 kg explosive charge.

DST-40 Quickstrike. A converted Mk-83 bomb as shown above. A magnetic bottom mine. Can be air-dropped by S-3, P-3, A-6, A-7 U.S. Navy aircraft as well as some 80 operational Air Force B-52D bombers. Theoretically, any U.S. Navy submarine could lay mines from its torpedo tubes.

DST-41 Quickstrike. A converted Mk-84 bomb as shown above. The Quickstrike series of mines contain a sophisticated solid state electronic sensor known as a target detection device. The EX-65 version of Quickstrike employs a new case which contains the largest amount of explosive of any mine in the U.S. Navy inventory.

Note: The Mk-82 bomb Quickstrikes are also known as EX-62. The Mk-83 bomb Quickstrikes are also known as EX-63. The Mk-84 bomb Quickstrikes are also known as EX-64. The new mine case Quickstrike is known as EX-65. The EX-64 may use a PBX explosive load.

The Quickstrike uses a target detecting device (TDD) EX-57 magnetic/acoustic and the EX-58 magnetic/acoustic/pressure. The TDD is modern state of the art electronic technology and allows for simple and fast assembly of the mine. Like the DSTs (Destructors), Quickstrike will use existing general purpose (GP) bombs except for the EX-65, which uses a new design mine case. The newest TDD 63/64 uses microprocessor technology. The delivery of all Quickstrike mines will be by aircraft carrier or land-based aircraft. Aircraft carriers of the U.S. Navy can stock a large amount of mines.

DST-115A A moored contact mine, surface-laid. Uses DST-801 firing mechanism. About $\frac{1}{2}$ meter in height and diameter. Approximately 60 kg. Arming delay 1 minute to 30 minutes; depending on water temperature and current velocity. Floods and sinks (still active) at 1, 12, 24, 48 or 72 hours. 1 to 50 meters bottom depth. Case is awash at surface. Expected life of 9 to 15 months. 10 meter minimum spacing.

DST-M117 300 kg aircraft-dropped mine. Mainly used by U.S. Air Force B-52D and F-4 aircraft.

Sweep Obstructor Mine Device. An anti-sweep moored contact device. Planted by aircraft. Target is sweep wires.

VEM (Versatile Exercise Mine). U.S. Navy has purchased a number of these from British Aerospace Ltd. in the U.K. for training and evaluation.

Nuclear Mines. Both the U.S.A. and U.S.S.R. are known to have nuclear sea mine capability.

Aircraft Mine-Carrying Capabilities

Table 3.1 lists various Naval aircraft of both fixed-wing and folding-wing types used from U.S. aircraft carriers that can carry and lay mines.

Table 3.2 lists various U.S. Air Force aircraft that can carry and lay mines. Notice that bombers, attack aircraft, fighter aircraft and cargo aircraft are capable of laying mines.

Submarine Mine-Carrying Capabilities

Table 3.3 is representative of the U.S. Navy submarine types and a typical mine load of three standard current mines.

U.S. Navy Aircraft Mine-Carrying Capabilities

Type Mine	*Aircraft Type*							
	A4	A6	A7	AV-8	P3	S3	F14	F4
Mine Mk25	1	5	6		7		2	
Mine Mk52	3	5	6		11	2		
Mine Mk55/56	1	5	6		7	2		
Mine Mk60	1	5	6		6			
Mine Mk65	1	5	6		6	2	4	1
Mine Mk64	1	5	6		6	2	4	1
Mine Mk62	14	28	20	5	18	6	14	18
DST Mk36	14	28	24	5	18	10		18
DST Mk36 Restr.	14	28	24	5	18	6		18
DST Mk40	7	10	10	2	8	2		8
DST Mk40 Restr.	5	10	4		8	2		8
DST Mk41	1	5	6		6	2		
DST Mk41 Restr.		5	6					

Note: There will be a variation of the above loads based on mission range, weather, etc.

TABLE 3.1

U.S. Air Force Aircraft Mine-Carrying Capabilities

Type Mine	Aircraft Type								
	B52D	A7D	A10A	F4	F16A/B	FB111	C130	C141	C5
Mine Mk52	18 (18)*			8			48		
Mine Mk55	20 (8)			4			20	29	108
Mine Mk56	12 (8)			4			24	29	108
Mine Mk60	10 (8)						24	29	108
Mine Mk65	18 (8)								
DST Mk36	60 (24)	20	20	22	14	**20-26-48			
DST Mk40	25								
DST M-117	36 (24)			10					

*Internal carriage ( )
**Dependent on wing sweep of 72, 55 or 26 degrees of the FB-111 aircraft.
Note: There will be some variation of the above loads based on mission range, weather, etc.

TABLE 3.2

Submarine Mine Loads
U.S. Navy

Submarine Type	Mine Mk 57	Mine Mk 60	Mine EX 67
SS-567	42	23	0
SS-572	28	28	0
SS-576	35	20	0
SSN-578	28	16	0
SSN-580	34	34	0
SSN-585	37	28	22
SSN-588	37	28	22
SSN-594 Less SSN-613, 614 & 516	46	24	24
SSN-597	26	14	0
SSN-613, 614, 615	54	54	54
SSN-671	54	54	54
SSN-685	54	54	54

Note: Some older types are shown here only for comparison purposes.

TABLE 3.3

U.S.S.R. Mines

The Soviets have been very innovative in some of their mine developments. Even in the days of the Czars, Russian technicians and engineers deeloped new designs and used novel approaches.

Soviet mines are broken down into certain categories. Some of these are listed below, with their designators.

AMD	Air, Bottom, Magnetic and Seismic
KMD	Surface, Bottom, Magnetic
MKB	Surface, Moored, Contact
MAG	Surface, Moored, Contact
KB KRAB	Surface, Moored Influence
RVM	Submarine, Moored, Passive and Active Acoustic
ARVM	Air, Moored, Passive and Active Acoustic
UEP	Submarine, Moored, Underwater Electrical Potential

The Soviet Navy has been placing renewed emphasis on mine warfare. This has been seen in the provision of specialized minelaying equipment on the "Kilo" class conventional submarine and by the increased numbers of airborne minelaying missions in Soviet exercises since the late 1970s. The Soviets still maintain considerable stockpiles of older mines. The specifications and designations of the major Soviet mines will be addressed in this section.

Note: On a typical mine designation such as M-12, the 12 indicated design promulgated in 1912. Also a mine designation of ADM-500 means that 500 kg of explosive is used. Contact mines were used during the Russian-Japanese War in 1904. Both Russia and Japan used them. After the war, Russian developments proceeded.

Soviet Submarine-Laid Mines

1917—The PL-100—Three submarine minelayers were in service with these.

1924—The PL-150 (with 150 kg of explosive filler) entered the fleet. This mine had an anchor and the assembly had neutral buoyancy. The trim of the submarine was not disturbed when this mine was moved aft in the mine-launching apparatus. This mine had reliable safety devices.

1927—New Leninets class submarine minelayers L-3-initiated. The L-3 laid some mines in August 1942 during WW II. When launched through a tube, the seventh mine case separated from the anchor. The anchor jammed in the tube. The mine was towed behind the submarine for 30 minutes and did not detonate because the safety devices had been activated.

WW II—PLt-G UEP mines were planted from K-class submarines. Had contact firing mechanisms. Were planted at pre-set depth below the surface. Used mainly in offensive minefields. Designer was Engineer Captain 3rd Rank N. Fedorov. UEP designator means "Underwater Electric Potential."

Soviet Aircraft-Laid Mines and Bottom Mines

1925—Development of aircraft mines began.

MAV — MAV mine is a moored contact-type mine, using parachute. Planted at preset depths below the surface.

AMG — The first non-parachute aircraft-laid mine. Resembles an aircraft bomb. Dispersion of the water entry points with the AMG is smaller than with parachute mine. This permits planting in ice channels and even breaking through some ice. Firing mechanism is contact-type with detonating galvanic horns. Planted at a predetermined depth beneath the surface.

MIRAB — Bottom mine with influence firing mechanism. Used during end of WW II. Designed for planting in rivers, channels, and harbors. Firing mechanism is of induction-type (first model developed in 1928).

ADM-500 — In December 1942 Navy put this bottom mine with an induction type firing mechanism into service. The -500 indicated 500 kg of explosive. These mines were armed with improved antisweep devices. L. Matveyev and I. Skvortsov were the design engineers of these mines.

The Soviets made comparative tests of their ADM-500 against the British and German influence mines in the Caspian Sea, to demonstrate the high sensitivity of their induction-type firing mechanisms. Under identical conditions with the same vessel, the British A-IV mine was actuated at a depth of 7 meters; the German mine with M-1 and M-2 magnetic firing mechanisms, at depths of 12 meters; and the Soviet AMD mine at depths of 30 meters.

Today, Soviet stockpiles of mines include some 100,000 moored contact mines and as many as 200,000 modern magnetic and acoustic magnetic induction mines.

Soviet Minesweeping

Toward the end of WW I, bow protector paravanes started to be employed for protecting warships against moored mines. Of the 3000 warships and merchantmen equipped with paravanes, not a single one was sunk by mines, although several of the ships cut 10 to 14 mines adrift on each voyage.

In 1926, the Soviets devised the MZ-26 sweep obstructor by Rear Admiral P. Kitkin. This quadruple-action sweep obstructor had a special magazine assembly in which were mounted four small buoys with buoy-ropes and explosive cartridges on them. When the buoy was set adrift by the sweep and towed, the buoy rope was twisted around the sweep element. The cartridge then came into contact with the sweeping element and detonated. When this occurred, the buoy rope of the sweep obstructor was also cut and the next small buoy would take up a ready position.

Soviet Naval Aircraft with a Minelaying Capability

Backfire B	Bear F
Badger A	Blinder A
Badger G	Mail (Mod)
Bison	May—their ASW aircraft

Soviet Surface Ships and Possible Mine-Carrying Capabilities

Class Ship	*Type Ship*	*Mine Types*—Total number can be carried of each mine: MKB, MKB-3, KB-KRAB, MAG	Improved KB KRAB, KMD-500, KB-UEP, Improved KB-UEP	KMD-II-500, KMD-IV-1000, KMD V-B	KMD-1000, KMD-II-1000, KMD-IV-1000
Sverdlov	CG, CL		82		56
Chapayev	CL		90		60
Kashin	DDG		18		12
Krivak-I, II	DD		20		14
Kotlin & Mod Kotlin	DD		56		40
Skoryy & Mod Skoryy	DD		50		36
Riga	FF		28		18
Petya-I, II	FFL		22		14
Grisha	PCE		18		12
T-43	PFR		16		10
S.0.1	PCS		10		8
Shershen	PT		4		—
Shmel	BMRL		6		4
Alesha	MSC		82 +22 upper deck stowage		60 +14
Vanya-I	MSC		8		6
Vanya-II	MSC		12		8
Zhenya	MSC		4		2
Sonya	MSC		8		6
T-43	MSF		16		10
T-58	MSF		24		16
Yurka	MSF		10		6
Natya	MSF		10		6
Sasha	MSM		10		6
Udaloyy	Guided-missile cruiser				

Note: See Figure 3.55 for stern view showing mine rails for launching surface launched mines.

Soviet Submarines with Torpedo Tubes and Mine Load Capability (Estimate)

Submarine Type	*Total Torpedo Tubes*	*Mine Load*
Quebec	4	16
Whiskey SS	6	28
Zulu SS	10	44
Romeo SS	8	28
Foxtrot SS	10	44
Bravo SS	4	8
Tango SS	10	44
Whiskey SSG	6	12
Juliett SSG	6	12
Golf-I, II, III, IV	6	12
November SSN	8	40
Echo SSN	8	18
Victor-I, II SSN	6	36
Alfa SSN	4	24
Echo II SSGN	8	16
Charlie-I, II SSGN	6	24
Papa SSGN	6	24
Hotel-II, III SSBN	8	16
Yankee-I, II	6	24
Delta-I, II, III	6	24

Summary of Soviet (and Russian) Mine Types

Note: This is not necessarily a complete listing but is representative of the older and more recent Soviet mines. Follows an approximate chronological listing.

Year	*Designation*	*Type*	*Firing Mechanism*	*Case Depth*	*Bottom Depth*	*Explosive Kg*	*Layer*	*Comment/ Minimum Spacing—mtrs*
1890	Connected	Moored	Contact	Similar to early experim.			Ship	
1904	–	Moored	Contact	Used Russo-Jap War.			Ship	Some floated open ocean
1908	M-08	Moored	Contact	6 m	110 m	115 kg	Ship	37 m
1910	P-13		Contact	Automatic floating			Ship	
1912	M-12	Moored	Contact	6 m	147 m	115 kg	Ship	30 m
1912	PL-100	Moored	Contact			100 kg	Submarine	
19--	S Mine	Moored	Contact	Used against small warships				
1916	Rybka	Moored	Contact	Used in ASW barriers WWI			Ship	
1916	M-16	Moored	Contact	6 m	365 m	115 kg	Ship	37 m deep mine
1924	Air Mine	By Kovalevskiy. Averin Influence Mine Exploder.						
1924	PL-150	Moored	Contact			150 kg	Submarine	
1925	MAV	Moored	Contact				Air	

Note: First development of air-dropped parachute mine. Planted at preset depths below the surface.

1926	M-26	Moored	Contact	6 m	139 m	240 kg	Ship	55 m

Note: Firing mechanism, depth mechanism, detonator, sinking device and anchor release gear were same as 1912 model. Had anti-sweeping devices—perhaps first in world. Mine case was mounted horizontally on the carriage. This lowered center of gravity, increasing dynamic stability and simplified mine preparation and sea launching. The high stability of the M-26 allowed launching even from patrol boats. During WW II, torpedo boats of the Red Banner Baltic Fleet planted them. Also were used in Black Sea, Barents Sea, Japan and Okhotsk Seas from various warships.

1930	AMG	Moored	Contact	Galvanic horns			Air with no chute—was free fall	
1931	KB-1	Moored	Contact	10 m	272 m	229 kg	Ship	42 m

Note: Designed for use against surface warships and submarines, both surfaced and at periscope depths. Firing mechanism is of galvanic-contact type. Depth for planting is about twice the maximum depth to which the M-26 could be planted with an explosive load of the same weight. This mine had protective shields. It made it possible to plant mines in small ice floes and blocks. Apparently only the Soviets used this automatically discarded hood.

The Soviets used steel antennas (steel wire rope). The difficulty in their use was that they appear to be a source of eddy currents in sea water, which brings about an actuation of the firing mechanisms in the absence of a target vessel. Self-actuation is also possible during rupture due to corrosion, and antenna polarizations during sea disturbance or during the detonation of an adjacent mine.

1933	MKB	Moored	Contact	10 m	272 m	229 kg	Ship	42 m
1934	MKB-3	Moored	Contact	10 m	272 m	200 kg	Ship	35 m

Note: Standard Soviet contact mine. Possibly 30,000 in stockpile. Ellipsoid shape. No ship count provisions.

1938	MIRAB	Bottom	Influence	10 m		66 kg	Air/Ship	60 m

Note: Soviet induction influence mines were far superior to German magnetic and British induction firing mechanisms.

1940	MAG	Moored	Contact	80 m	457 m	229 kg	Ship	35 m

Note: Primary target was submarines. Ellipsoid shape, about 10,000 in stockpile. Sometimes chemical horns of MKB are used. Optional upper antenna supported by a float 50 m long. Lower antenna is a little longer.

1940	PLT-G	Moored	Contact	10 m	137 m	229 kg	Sub	80 m Detonate range 30 m
1940	R	Moored	Contact	2 m	35 m	10 kg	Ship	15 m
1940	R-1	Moored	Contact	2 m	35 m	40 kg	Ship	20 m
1940	MYaM	Moored	Contact	3 m	50 m	20 kg	Ship	20 m
1941	PLT-3	Moored	Contact	10 m	128 m	100 kg	Sub	30 m
1942	AGSB	Moored	Contact		500 m	229 kg		

Note: This is a deep-laid mine with two steel antennas, each 35 m long. The problem of interference was solved by the use of a two-pulse operating circuit.

1943	AMD-500	Bottom	Influence	25 m	25 m	300 kg	Ship/Sub/Air	68 m

Note: Air version uses parachute. Cylinder shape. Approximately 18,000 in stockpile. Magnetic influence-type mechanism. KMD indicates surface-laid.

1943	AMD-1000	Bottom	Influence	54 m	30 m	700 kg	Ship/Sub/Air	137 m
1943	AMAG-1	Moored	Contact	9 m	100 m	259 kg	Air	45 m

Note: Free-fall mine with shallow ice penetration capability. Ogival nose and bomb-type fins are added. Ellipsoid-shape case.

1943	Allied	Moored	Acoustic Influence	Lend-Lease from Allies				
1943	A-IV	Moored	Magnetic Influence	Lend-Lease from Allies			Ship	

1944	UEP	Moored	Contact Influence	Electrical Potential First of the series 487 m		226 kg	Sub	30 m range
1945	M-1, M-2	Moored	Influence Magnetic				Ship	Captured from Germans 1945
1949	KB-KRAB	Moored	Influence Acoustic	20 m	272 m	229 kg	Ship	42 m

Note: Uses double passive acoustic firing mechanism. Ellipsoid shape. "Krab" acoustic device is inserted into a standard MKB mine case. Two passive frequencies may be 22.5 KHz or 30-40 KHz.

1949	MKD	Bottom	Influence	54 m		782 kg	Ship/Sub	42 m
1950	BPM-2	Limpet type				4 kg	Swimmer	
1950	MZ-26	Moored	Contact	34 m	45 m	2 kg	Ship	

Note: This is a sweep-obstructor type mine.

1950	AMD-II-500	Bottom	Acoustic Magnetic Induction	50 m	50 m	300 kg	Ship/Sub/Air	
1950	AMD-II-1000	Bottom	Acoustic Magnetic Induction	50 m	50 m	700 kg	Ship/Sub/Air	

Note: Surface launched versions are KMD-500 and KMD-1000. They use strap-on dolly. Air-dropped version uses parachute. Both versions could be submarine-launched.

1951	KB-UEP	Moored	Electrical UEP Firing	40 m	450 m	240 kg	Ship	

Note: Underwater Electric Potential (UEP) detected by upper and lower electrodes. Has 2 random looks firing logic. Primary target is the submarine. Has ellipsoid-shaped body. Possibly 10,000 in stockpile.

1954	AMD-IV-500 AMD-IV-1000	Bottom	Pressure Acoustic Magnetic	50 m	50 m	300 kg 700 kg	Ship/Sub/Air	

Note: These mines have ship-count provision of 1 to 21. Long delay arming is 12 hrs to 10 days. Firing logic takes 3 looks. Has anti-countermine provisions. Cylinder-shaped case. Possibly 35,000 in inventory. During the Korean War the landing of 50,000 United Nations troops at Wonsan had to be delayed for eight days. It was discovered that a force of North Korean junks and sampans had laid 3,500 Soviet magnetic influence mines. The magnetic exploder mechanism in them could not be neutralized by the standard "degaussing" or neutralization of hull magnetism. Soviet technicians no doubt helped in planting and adjusting these mines.

1960	Improved KRAB	Moored	Passive Active Acoustic	150 m	600 m	250 kg	Ship	

Note: Has improved case and warhead. Passive frequency could be about 200-250 Hz with active echo ranger at 30-40 KHz. Has ellipsoid case. Ship count of about 1 to 21. Delay for arming is up to 21 days. Firing logic has 2 looks. Sterilization can be set to 370 days. Possibly 10,000 in inventory.

1960	Improved KB-UEP	Moored	UEP Firing	150 m	600 m	250 kg	Ship	

Note: May use longer electrodes for sensors. Has ellipsoid case. Ship count of 1 to 21. Delay arming is up to 21 days. Sensor has 2 looks at random. Sterilization is adjustable to 370 days.

1969 AMD-P-V Bottom Seismic Magnetic Gradiometer 80 m 80 m 1000 kg Ship/Sub/Air

Note: Possibly has a remote on-off capability. Has cylinder case shape. Ship count of 1 to 21, delay adjustable to 30 days, sterilization adjustable to 370 days and has anti-countermine feature. Possibly 12,000 in inventory.

1969 RVM, A-RVM, G-RVM Moored Passive Active Acoustic Rising 300 m 750 m 220 kg 200 kg Ship/Sub/Air

Note: Acoustic rising mine detects presence of target in 46 KHz passive mode. Determines target range when directly over mine in 158 KHz active mode. The rocket motor then propels the mine to detonate at target depth. May be 15,000 in inventory. The GRVM is a guided version for use against submarines and aircraft carriers. Anchor can be up to 1000 m deep, minimum of 60 m. Case can be up to 400 m deep with minimum of 60 m. Possible attack range to target is 600 m. Firing mechanism takes 2 looks at target, then raises mine to target. Has no ship count. Long-delay arming is up to 21 days. Sterilization is adjustable to 370 days. Frequency of active mode may be adjustable from the 158 KHz ping range. Cylinder case.

1969 UEP Moored UEP 100-200 m 600 m 150 kg Two charges Ship/Sub

Note: This UEP mine deploys into an 80-m-long array. The case, depths are various with float, explosive and firing mechanism separate. Has an inhibit option as a counter-countermeasure. Two warheads are carried. Has cylinder case. Approximately 10,000 in inventory. The UEP is sometimes called a "Sectionalized" mine because of the separation of the three case sections.

1970 AMD-S Remote-Control Minefield. Can be actuated from remote location.

1980 RMZ Anti-Helicopter Mine. A mine that is probably actuated by the sounds from the minesweeping helicopters. The intent would be to explode the mine with a force that could destroy or disable the helicopter. This is essentially an anti-sweeper mine. Weight on the order of 700 kg, with payload approximately half that. This mine is about 2 m long.

1982 SLMM Soviet Submarine Standoff Mine. Can be launched from their submarines. Has a range of about 30 km.

1970-80 It is expected that the Soviet Navy has nuclear mine capability.

Note: Table 3.4 (on the next page) lists some current Soviet mines. Extracted from a Soviet Journal. See Figure 3.56.

1980 99501 NG63 Soviet-made mine recovered by the Royal Navy near the Suez Canal. 3 m long, 533 mm (21 inches) in diameter. Had acoustic, pressure and magnetic firing mechanisms. Possibly 1981 manufacture date. See Figure 3.57.

19-- YaRM 3 kg charge

Yam 20 kg charge

Note: These are smaller-sized contact mines ready for deployment in shallow waters, conventionally moored.

The Soviet mines vary in size and charge, some up to 1000 kg. The equivalent to the U.S. Navy "Captor" are the Soviet "Cluster Bay" and "Cluster Gulf" NATO designations. They are otherwise known as "Rising Mines." Their speed when released is higher than that of a normal torpedo since a rocket motor supplies the propulsive power. They need to have only limited range since they are designed to attack only in the vertical plane, the anchorage point. The fuel load can therefore be devoted entirely to high-speed propulsion. The speed at which the weapon approaches the target allows little time for defensive reaction or use of a countermeasure.

Наименование мин	Масса, кг: общая / заряда ВВ	Габариты: длина, м / диаметр, мм	Глубина постановки, м
МСС 23С	350 / 500	2,4 / 533	...
МСТ 15	1500 / 1000	[illegible] / 1200	...
«Манта»	220 / 100	380 / 980	2—100
MR-80C	500 / 400	1740 / 533	3—300
VP-700	700—1700 / ...	... / 533	...
U53M-600	700 / 600	2750 / 533	10—120
TAR-6	1456 / ...	... / ...	45—300
TAR-16	1500 / ...	... / ...	120—500
«Си Урчин»	... / 750	1,0 / 533	40—90
«Стоунфиш»	205—700 / ...	1,0 / 533	30—200 (против ПЛ)

TABLE 3.4

The Soviets are well advanced in mine development and the Soviet naval fleets have a considerable mine warfare potential, in terms of both quality and quantity. They are ready for both defensive and offensive mine warfare.

The foregoing listing of Soviet mines, old and new, is rather impressive. It shows the great emphasis that they have placed on mines and mine warfare. The U.S.A. could possibly learn a lesson from this for the future.

Thailand

Thai Navy has ordered the U.K. British Aerospace's Naval and Electronic Systems Division VEM (Versatile Exercise Mine). These will be used for training and evaluation of the Thai Navy's mine countermeasures operations.

General Mine Information—New Development

See Figure 3.58 for a picture of a novel Passive Naval Mine System. Tek-Sea, an engineering company with headquarters at Lugano, Switzerland and operating offices in Milan, Munich and London has developed a passive anti-ship mine system referred to as a "Telemine." It can be deployed from ships, submarines or helicopters on the sea bed to depths of 200 meters. It will lie dormant for up to six months or until activated by acoustic signals from up to 40 km away or by electromagnetic signals from up to 500 meters away. Activation by these signals will cause ballast to be released, allowing the Telemine to rise to just below the surface. Remote-controlling stations may then feed it an autopilot program of action, depending on the objective.

The Telemine carries a 300 kg warhead. It is powered by a battery-operated 6 hp motor driving a small propeller like that on a torpedo.

Bottom Mine

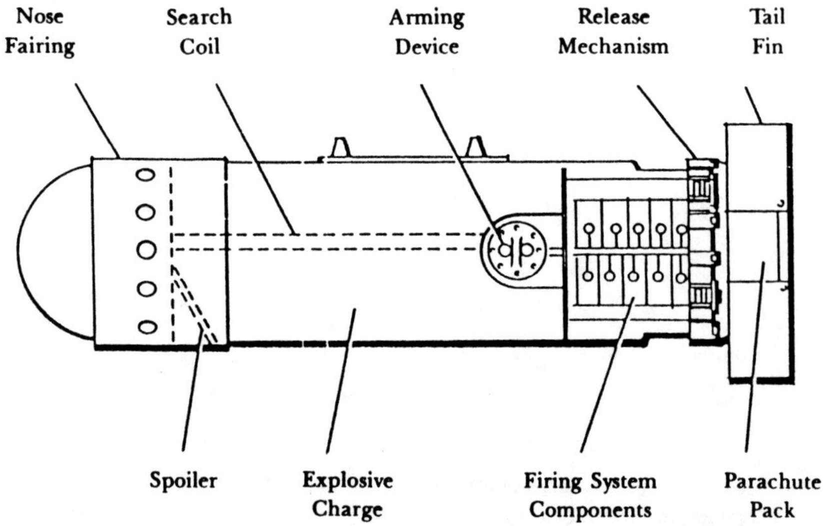

Mine MK 55

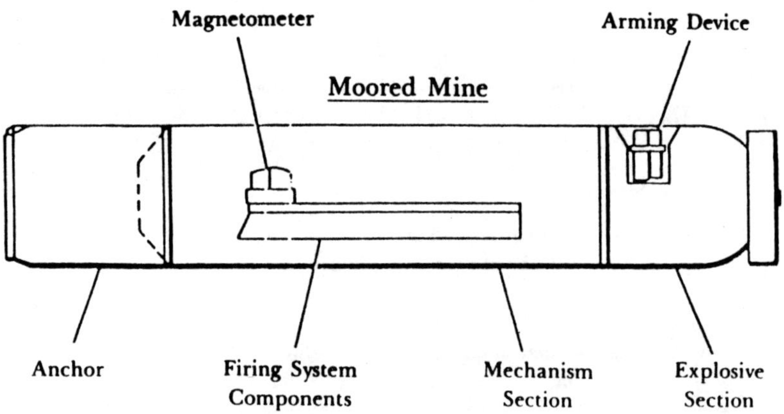

Figure 3.1. Typical Mine Components.

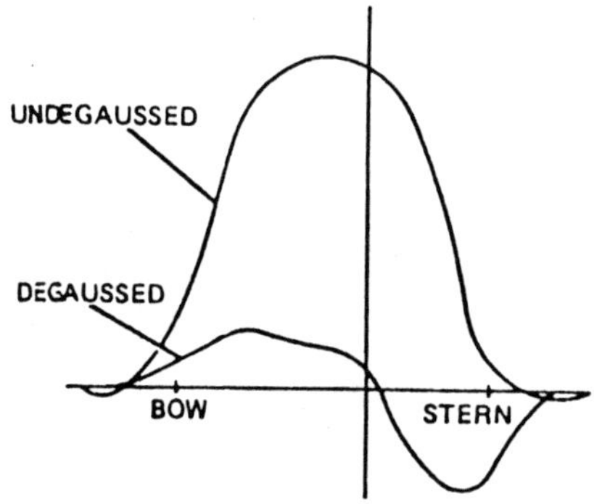

Figure 3.2. Typical Magnetic Signature.

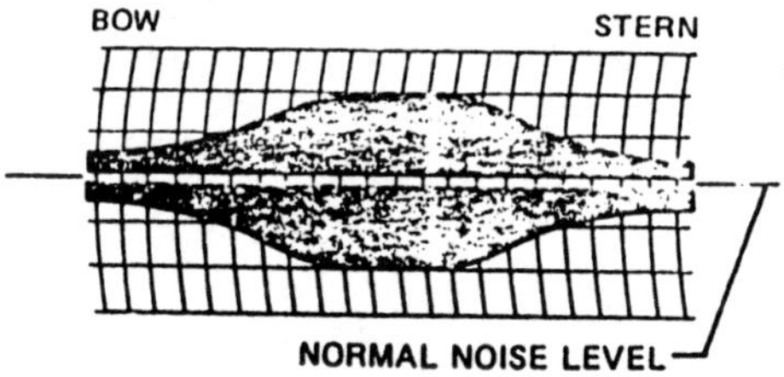

Figure 3.3. Typical Acoustic Signature.

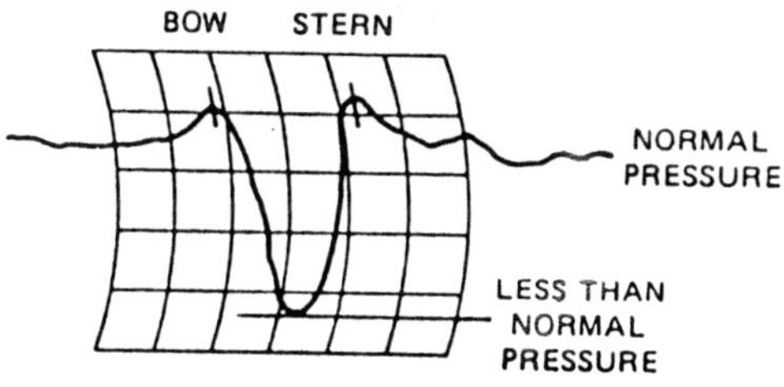

Figure 3.4. Typical Pressure Signature.

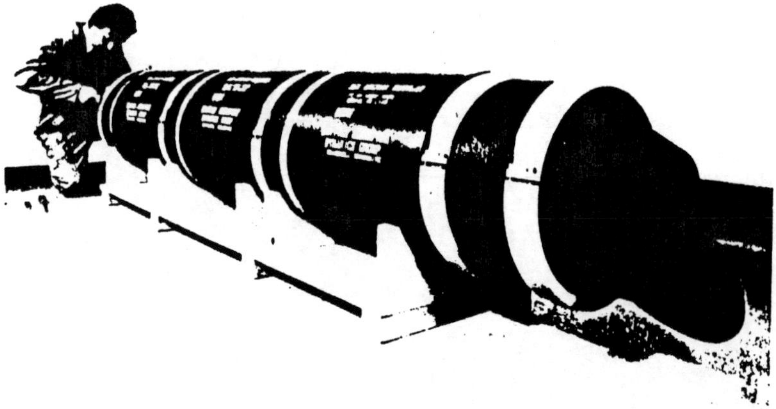

Figure 3.5. British Aerospace's new "Sea Urchin" ground mine, a multi-influence mine which can be laid by ship, submarine or aircraft.

Figure 3.6. The British "Dragonfish" ground mine. Influence type produced by Marconi Ltd.

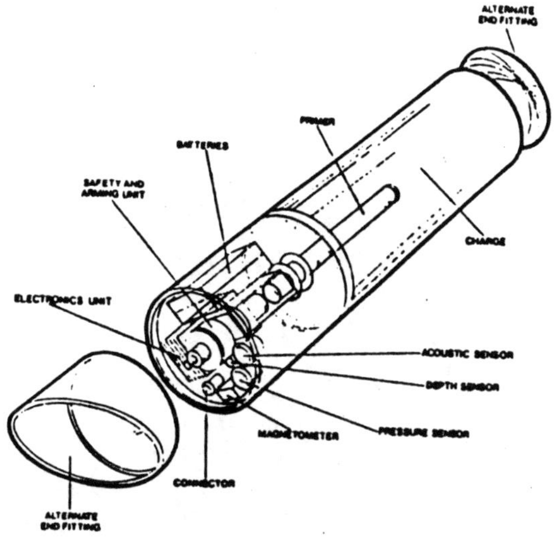

Figure 3.7. A schematic of the "Sea Urchin" produced by the British company BAe Dynamics.

Figure 3.8. A British Hovercraft that—because of its method of travel—is clear of the water and therefore is little affected by an underwater explosion.

Figure 3.9. British "Stonefish" modular ground mine for use against ships and submarines. Mines being loaded into a transport aircraft. This mine is produced by Marconi Ltd.

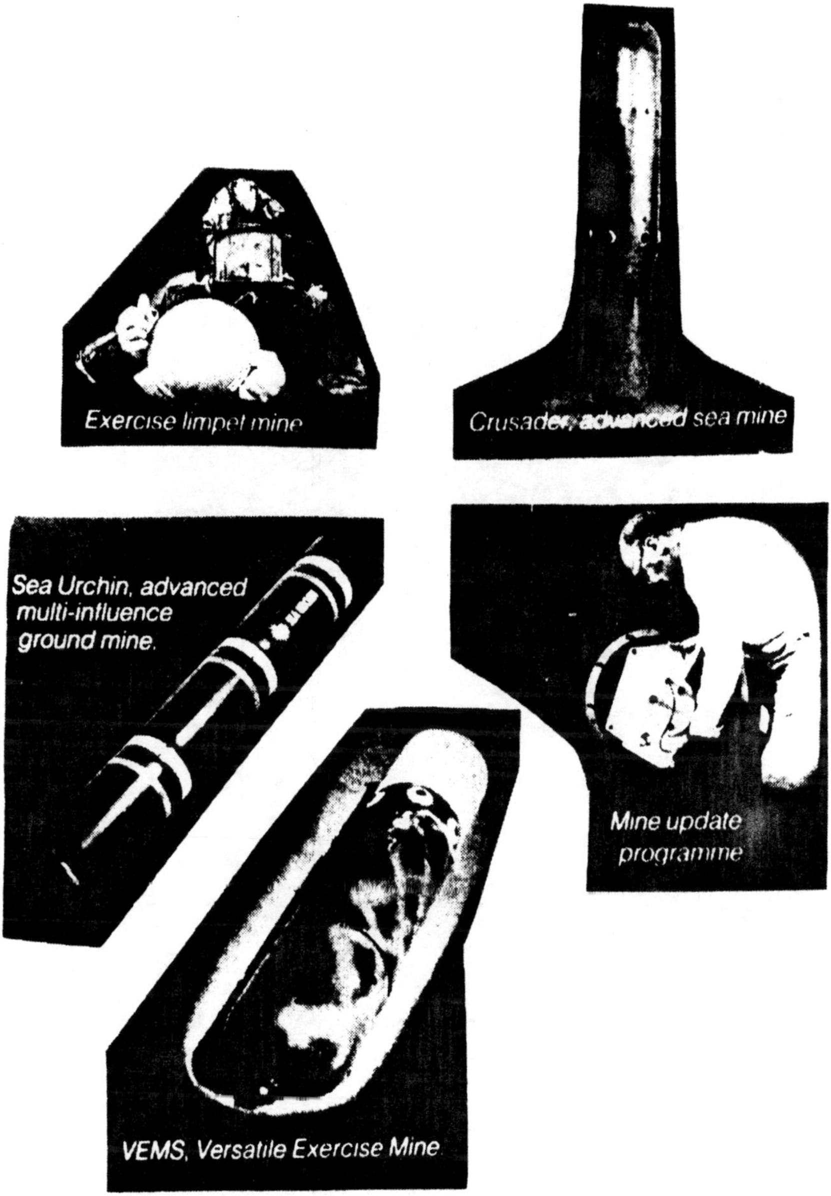

Figure 3.10. Various British Aerospace Ltd. mine programs.

Figure 3.11. Obsolete British practice mines on deck of "Hunt" Class MCMV.

Figure 3.12. French Navy moored influence mine MCT built by Thomson-CSF.

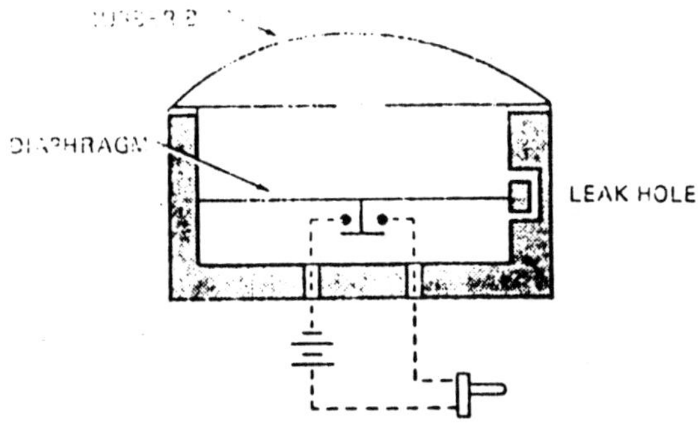

Figure 3.13. The famous German "Oyster" mine mechanism used to delay the Allied invasion of Europe in 1944. Used a simple method of detecting the pressure drop under a moving ship. A rubber bag exposed to the sea covered a space which was divided into two parts by means of a diaphragm. When the mine is laid,, the pressure on either side of the diaphragm is equalized through a leak-hole, and any slow variations of pressure due to swell or tide are taken care of in the same way. However, when a ship passes over the mine, the sudden reduction in pressure corresponding to a head of about two inches of water cannot be dealt with quickly enough by the leak-hole. The pressure on either side of the diaphragm becomes unbalanced, and it is sucked toward the rubber bag and so completes the firing circuit to the detonator.

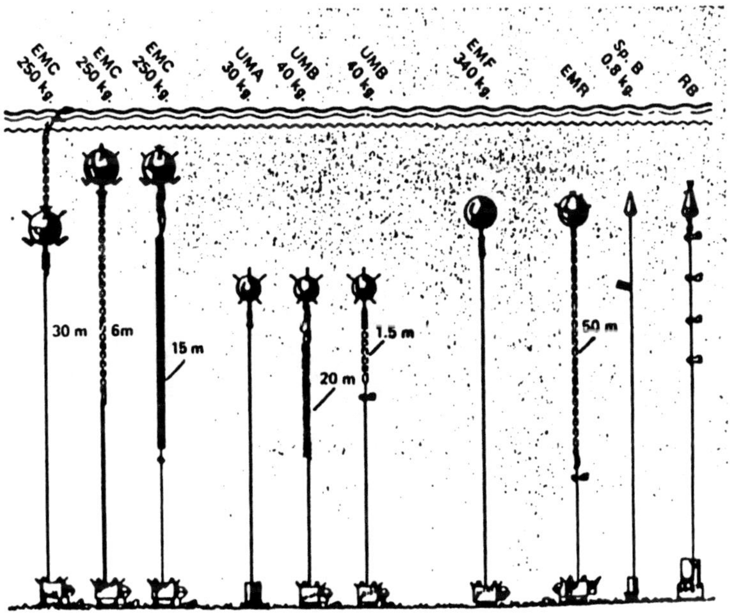

Figure 3.14. Primary mine types used by Germans in World War II. About 150,000 of them were laid.

	Days Port Closed	*Mines Laid*	*Ships Sunk (or Damaged)*
St. Johns	0	81	2
Halifax	1	60	1
Boston	0	10	0
New York	2	10	0
Baltimore	0	10	1
Norfolk	3	47	5
Wilmington, N.C.	6	0	0
Charleston, S. C.	16	36	0
Savannah	3	0	0
Jacksonville	3	12	0
San Juan	0	17	0
St. Lucia	6	6	2
Trinidad	0	14	0
Panama	0	15	0
Mississippi River Area	0	9	0
	40	327	11

Figure 3.15. German submarine minelaying in Western Atlantic, World War II.

Figure 3.16. East German mine warfare training.

Figure 3.17. East German and Polish sailors training on mines.

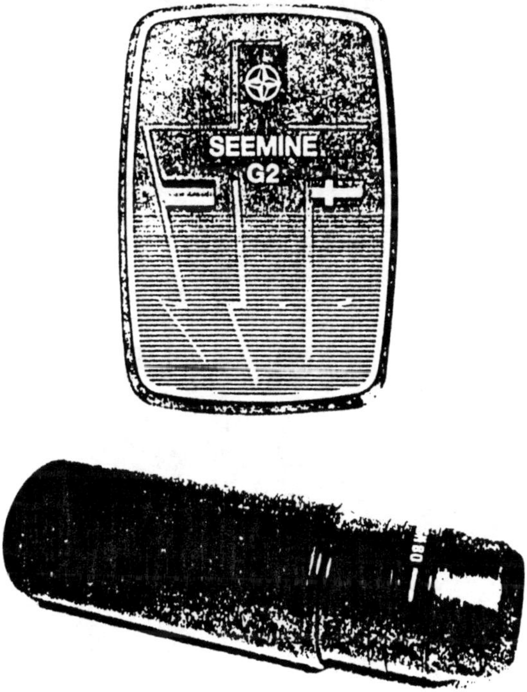

Figure 3.17.1. The German/Danish Naval G2 Seemine G2 operational configu...

Figure 3.17.2. The G2 mine in the practice configuration. Light buoys and light line are shown extended from housing.

Figure 3.17.3. German Navy G2 Ground Mine launching from test ship in Eckernforde, West Germany.

Figure 3.17.4. The G2 "Seemine" on Royal Danish Navy ship. This mine is a joint German-Danish program.

Figure 3.18. Italian Navy TAR 6 and TAR 16 mines produced by Whitehead Moto Fides. Can be laid in depths to 500 meters and will form a barrier at a maximum depth of 300 meters.

Figure 3.19. Typical exercise mines. Can be used for searching and hunting operations.

Figure 3.20. Italian Navy MRP advanced general purpose sea mine. Can operate at depths to 300 meters.

Figure 3.21. Italian Navy ground mine WP-900/1. Difficult to sweep. They are fitted with fuses that allow different permutations of ship counts to be applied.

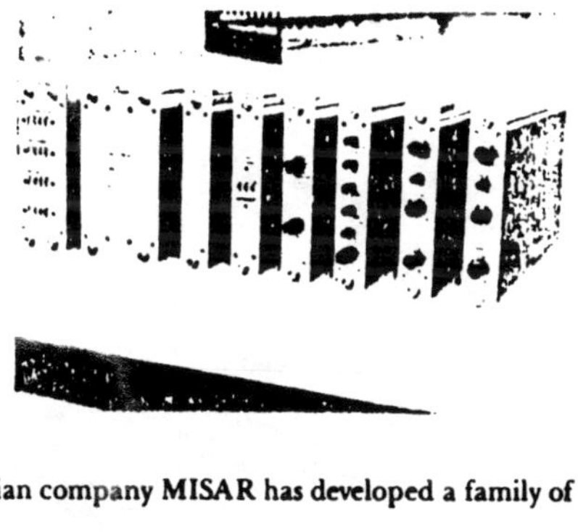

Figure 3.22. The Italian company MISAR has developed a family of fuses for use in their mines.

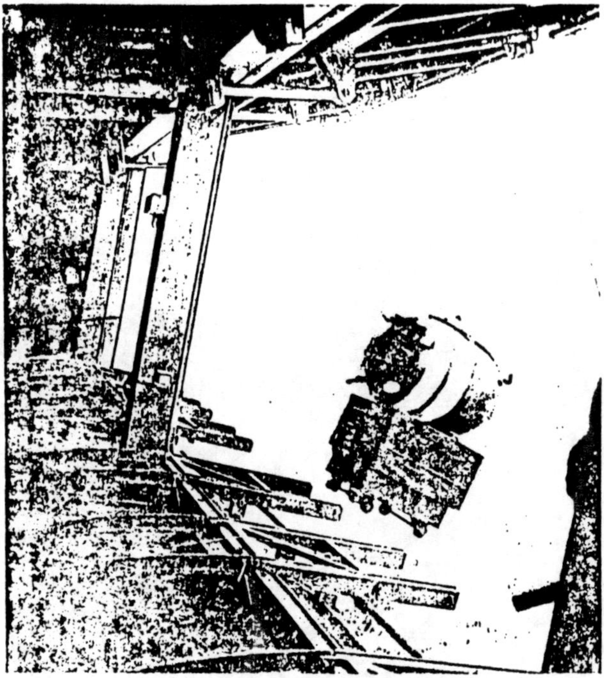

Figure 3.23. Launching a mine from rails on a surface ship.

Figure 3.24. Contact mines used by Iran in the Persian Gulf in September 1987.

Figure 3.25. Swedish Navy Bofors "Rockan" mine.

Figure 3.26. Royal Swedish Navy K11 Bofors MM180 moored mine.

Revolutionary Powder Keg Mine

Civil War Mine
Spar Torpedo

Civil War Mine
Singer Type

Civil War Mine
Brooks Type

Figure 3.27. Various U.S.A. and Confederate mines used in the past.

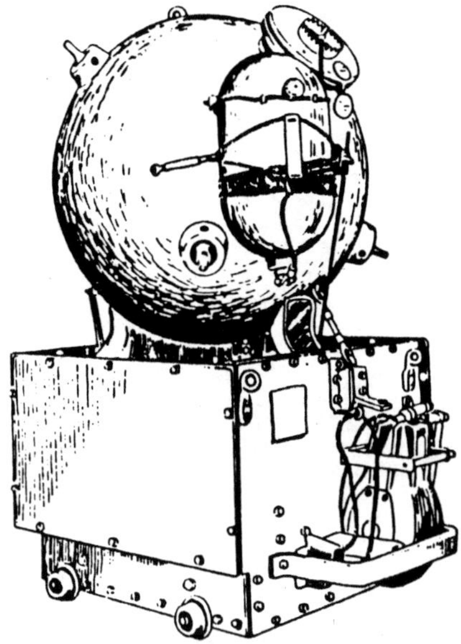

Figure 3.28. The U.S. Mark 6 mine used in North Sea Mine Barrage in 1918.

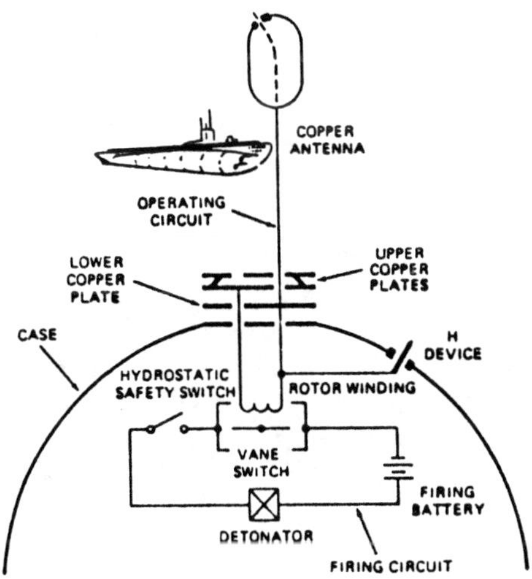

Figure 3.29. Diagram of the K-Device firing circuit used in the Mk-6 mine. Perhaps named for Knott Apparatus Co. by Mr. Browne of Salem, MA, who worked there.

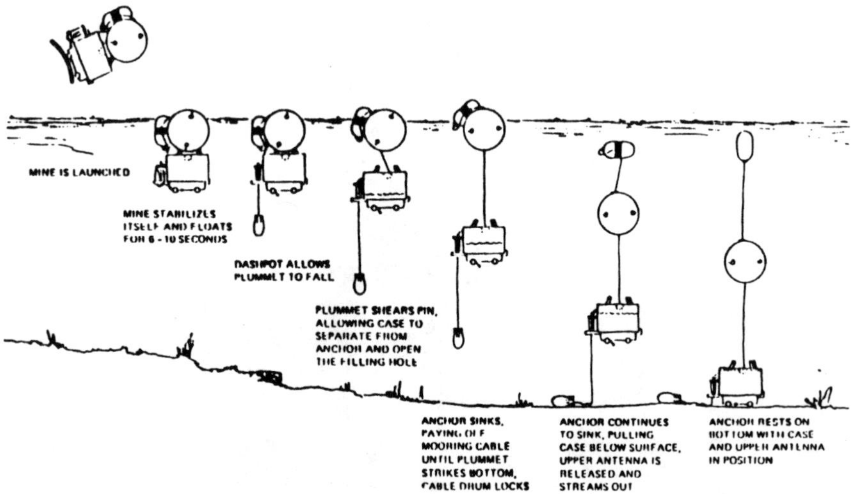

Figure 3.30. Mk-6 mine-planting sequence.

Figure 3.31. Mk-6 moored mines assembled in a ready issue store at Inverness, Northern Scotland. Used in Great Mine Barrage in North Sea in 1918.

Figure 3.32. Mine stockpile on Tinian, June 1945. These were dropped by B-29 bombers of U.S. Army Air Corps, during Operation Starvation against Japan.

Figure 3.33. U.S. Navy surface laid mines moored contact Mk-6 and Mk-16.

Figure 3.34. US Navy surface-laid mine bottom influence type Mk-18.

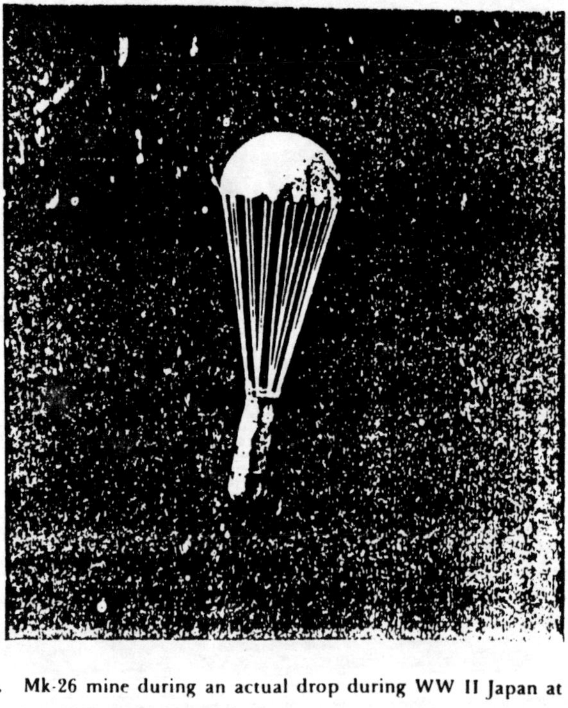

Figure 3.35. Mk-26 mine during an actual drop during WW II Japan at Chichi Jima.

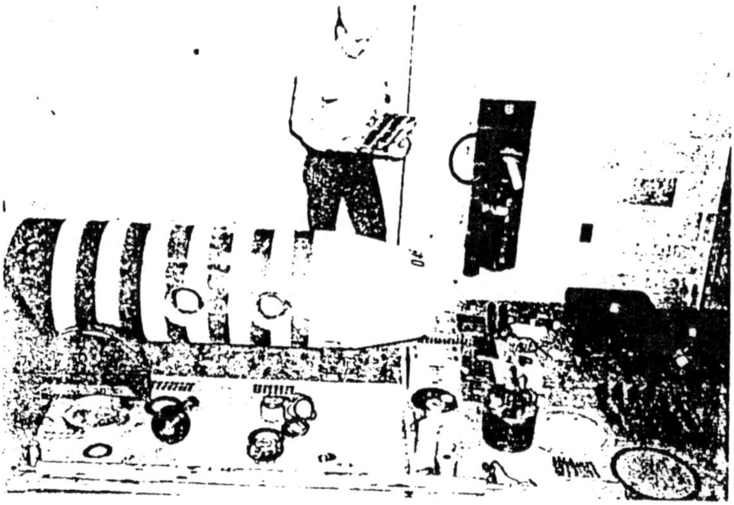

Figure 3.36. Mk-25 mine, showing components.

Figure 3.37. Mk-27 submarine-launched mobile mine.

Figure 3.38. Typical submarine-laid moored influence mines. Top, the Mk-12; bottom, the Mk-57.

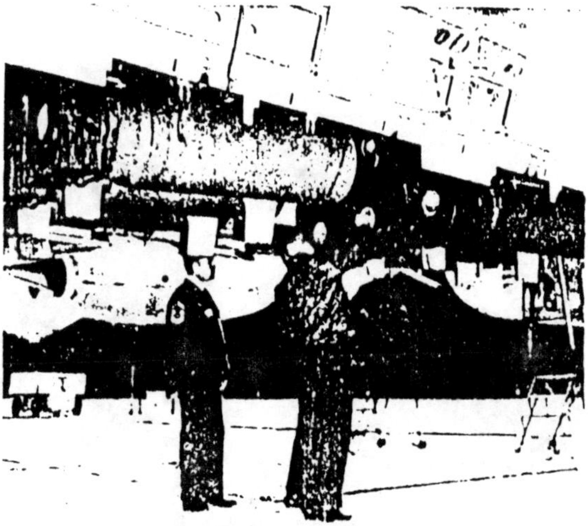

Figure 3.39. Loading mines in U.S. Air Force B-52 bomber.

Figure 3.40. U.S. Navy air-laid mine Mk-36 bottom influence mine.

Figure 3.41. U.S. Navy air-laid mines Mk-52 on left, Mk-55 on right, bottom-influence mines.

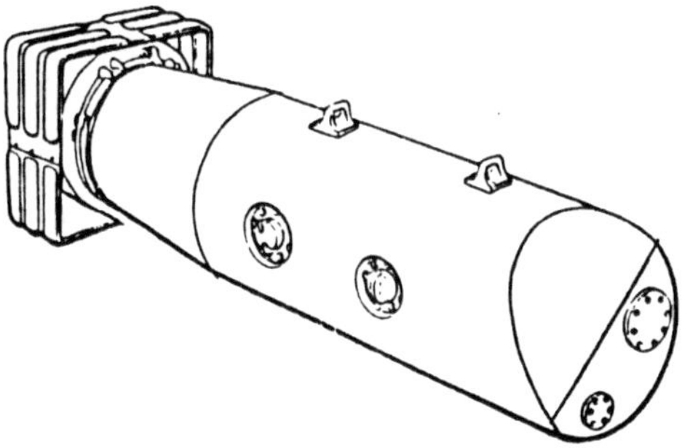

Figure 3.42. U.S. Navy Mk-25 mine. aircraft-laid magnetic bottom type.

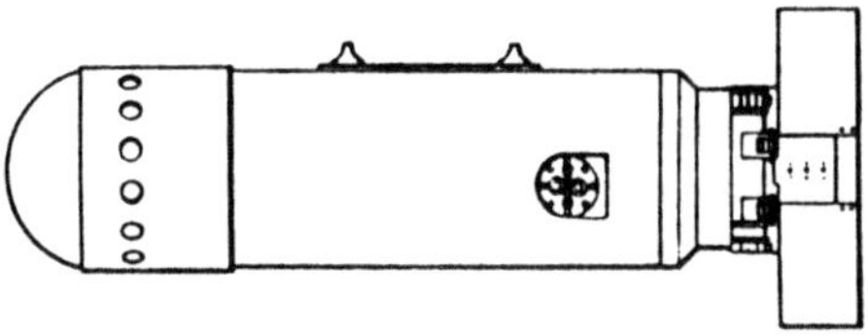

Figure 3.43. U.S. Navy Mk-52 mine. Aircraft-laid influence mine actuated on acoustics.

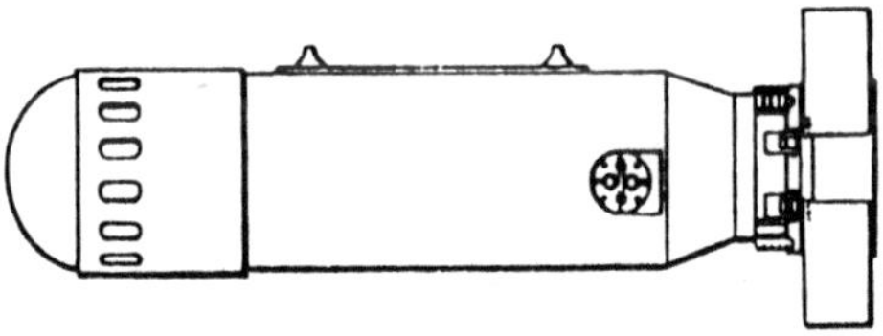

Figure 3.44. U.S. Navy Mk-55 mine. Aircraft-laid bottom type.

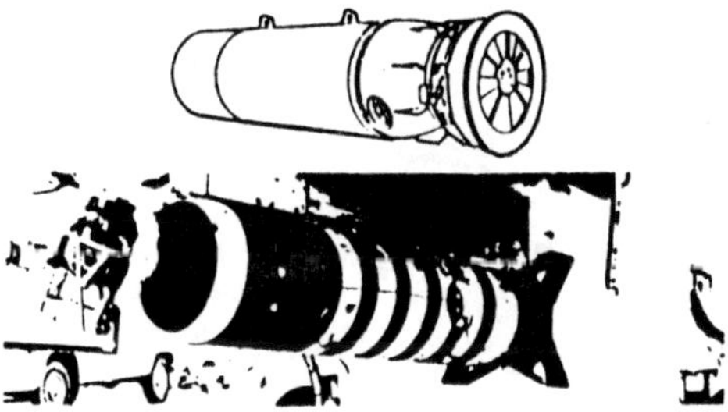

Figure 3.45. Two views of the U.S. Navy Mk-56 mine. Aircraft-laid. Uses magnetic firing mechanism. Bottom view shows installation on U.S. Navy A4B Skyhawk attack aircraft.

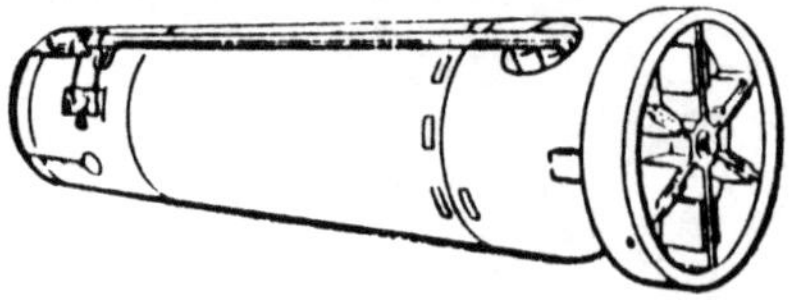

Figure 3.46. U.S. Navy Mk-57 mine. Submarine-laid. Has fiberglass case and steel anchor.

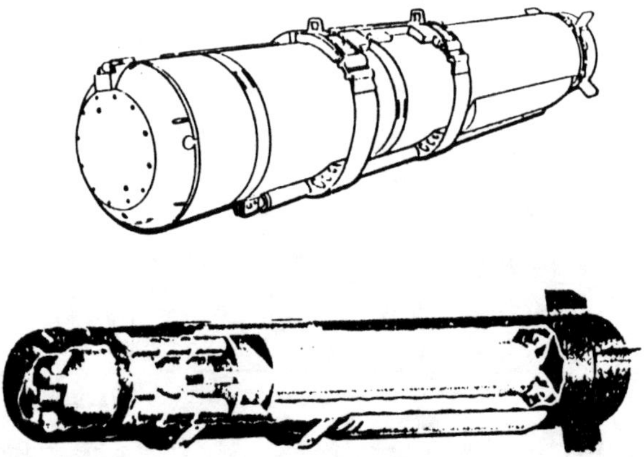

Figure 3.47. Two views of the U.S. Navy "Captor" Mk-60 deep-water mine. Weight is about 1,000 kg. The sensor releases a torpedo which then homes on a target.

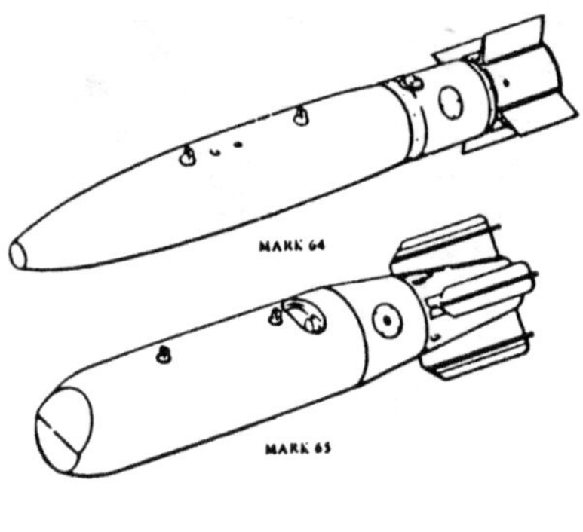

Figure 3.48. U.S. Navy Mk-64 and Mk-65 mines use variable influence target detection systems.

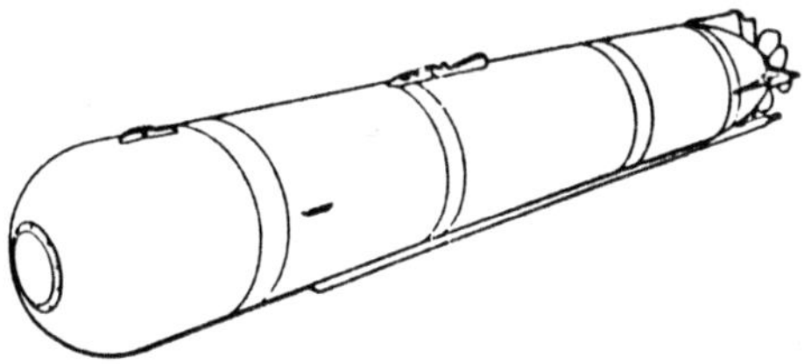

Figure 3.49. U.S. Navy Mk-67 multiple-influence bottom mine. Launched from submarine. Uses modified Mk-37 torpedo.

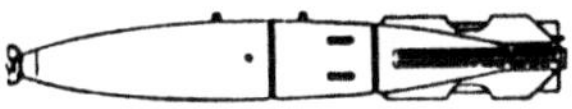

Figure 3.50. U.S. Navy Destructor (DST) Mk-36.

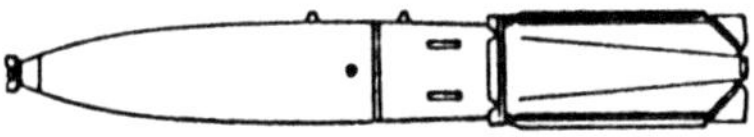

Figure 3.51. U.S. Navy Destructor mine (DST) Mk-40.

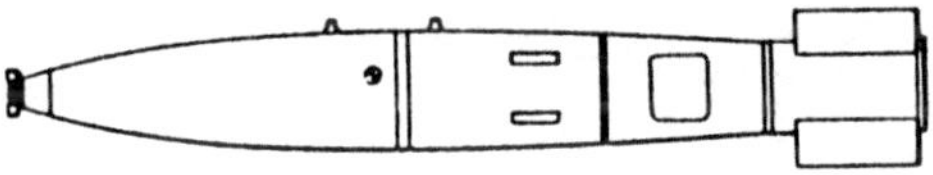

Figure 3.52. U.S. Navy Destructor mine (DST) Mk-41.

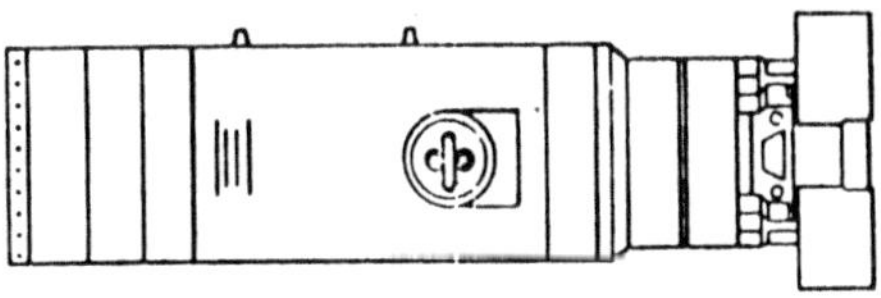

Figure 3.53. U.S. Navy actuation mine Mk-52. An exercise and training mine. A nondestructing version of the Mk-52 service mine.

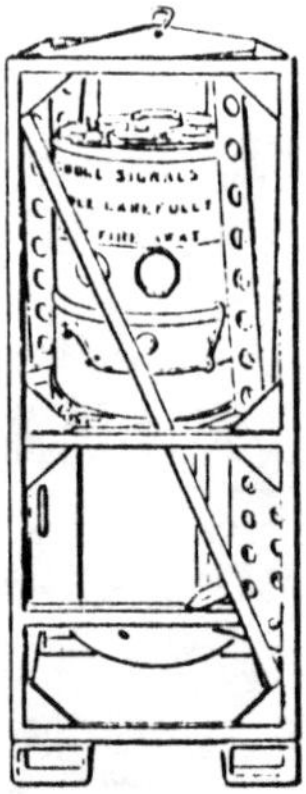

Figure 3.54. U.S. Navy actuation mine simulator Mk-61. An exercise and training device that simulates the response characteristics of a service mine.

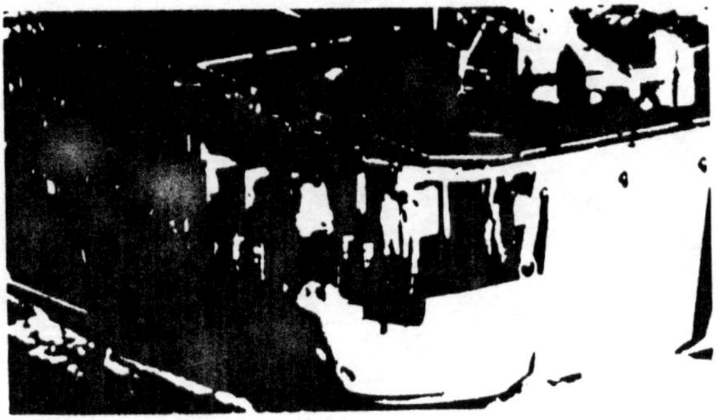

Figure 3.55. Soviet "Udaloyy" class ship fitted with mine rails on port side of stern.

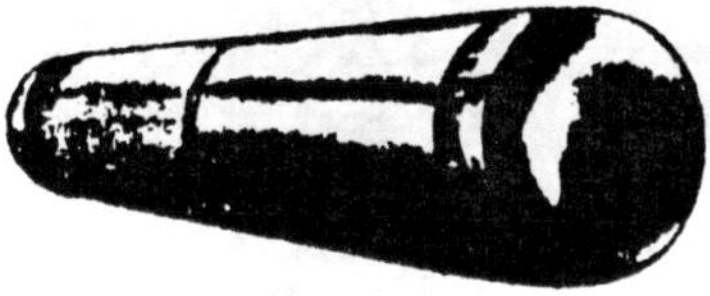

WP-900

TAR-6

TAR-16

MCC-23C

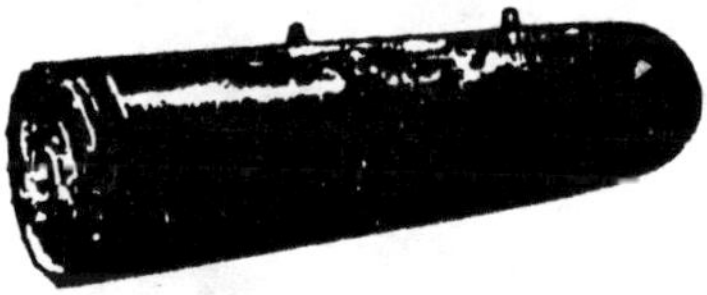

MR-80

Figure 3.56. Typical Soviet Navy mines.

Figure 3.57. Soviet-made mine instrumentation section recovered by Royal Navy near Sue Canal. 10 feet long, 21 inches in diameter. Marked 99501 NG63 with acoustic, pressure an na[illegible]tic detonators. Possible 1981 manufacture.

Figure 3.58. The new "Telemin developed by Tek-Sea Cc

CHAPTER 4

Minehunting/Mine Countermeasures Sonars

The sonars used to locate and classify naval mines are a natural follow-on to sonars used on surface ships for searching for submarines in the ocean depths. While surface ships surveillance sonars are usually low frequency in order to obtain the greatest range, sonars used for searching for mines are higher frequency.

Mine location sonars are divided into search and classification type. The search sonar is of lower frequency to obtain a longer range; the classification sonar uses the higher frequencies to provide greater detail on the object in question. Once an object is located by the search sonar, the classification sonar is used to determine whether the object is a mine (and details on the mine) or a rock, old barrel, refrigerator or similar debris that lies on the bottom of the area being searched.

In contemporary warfare, the sonar is only one part of the entire integrated mine warfare electronics package. This includes sonar, navigation equipment, plotting equipment, computers, data-gathering and recording devices, etc. Because the mine warfare vessel of today may be subject to being attacked by mines that can home on the minehunting vessel, some consideration must also be given to counter-countermeasures on the vessel. Many countries are also equipping their mine warfare vessels with anti-aircraft guns and missiles, as well as the electronic warfare receiver to detect electronic emissions. One can see why the modern minehunter is an expensive vessel.

The following listing of sonars is not complete, but is representative of the major types of sonar equipment used in the principal countries of the world that are involved in mine warfare. Some countries find it more practical and economical to purchase the sonars for their vessels from a country such as France that has a great deal of experience in the field and has high-quality sonars available on the open market. Other countries have obtained vessels from a major power, either by purchase or grant. These vessels normally come equipped with the sonars of the original vendor, such as the U.S.A. or the U.S.S.R.

Australia

MWS-80 produced by Atlas Elektronik of West Germany is installed in the Australian RAN "Bay" class vessels. This is a mobile van that can be moved from vessel to vessel. DQS-11H is a sonar also built by Atlas Elektronik and installed in the RAN "Bay" mine vessel. This is hull mounted but can be lowered from the hull. More detail on the MWS-80 and DSQS-11H is provided in the German sonar section.

Model 590 sonar has been selected by the RAN for their "Craft-of Opportunity" vessel program. This sonar is a side scan sonar transceiver built by the Klein Co. of U.S.A. Also included is a Model 595 graphic recorder.

France

Some of the most popular minehunting sonars are built by the French companies. There is considerable support from the French Navy in the development and marketing efforts for these sonars. The Direction des Constructions Navales (DCN) is involved in the vessel constructions.

Because some of the first French minehunting vessels and sonars after World War II were of U.S.A. origin, the UQS-1D U.S. Navy sonar was in use in these vessels.

The French Navy system called "Skubemor" includes a sonar and the PAP-104 wire-guided remote minehunting submersible. The PAP-104 has a version that uses a high-frequency classification sonar in the forward section. One of these sonars on the PAP-104 can be the West German AIS-11, built by Atlas Elektronik.

DUBM-20 Minehunting sonar. Consists of one sonar for detection, and another for classification. The classification sonar is of very high frequency for clarification and bringing out the details on the object being viewed, be it a mine or other underwater object. Manufacturer is Thomson-CSF and DCN. (See Fig. 4.1.)

DUBM-21/21A/21B minehunting sonar. One of the series of new French Navy sonars. Built by Thomson-CSF. The DUBM-21B double hull sonar detects and classifies ground and moored mines. Has search and classification sonars. Is used on the "Tripartite" class minehunters of France, Belgium and Netherlands. These sonars are sometimes called TSM-2021 and TSM-2022 "IBIS" systems. (See Figs. 4.2 through 4.4.)

TSM-2022 system is hull mounted and can be retracted into a well in the ship's hull. The array is 1.5 meters in diameter and a pivoting system allows it to be withdrawn into a well of only $\frac{3}{4}$ meter diameter. Weight of the system is $1\frac{1}{2}$ metric tons. The sonar is domeless so that dome losses, usually about 3 dB, are eliminated, with a consequent increase in range. Electrical rather than hydraulic stabilization is used, which gives a reduction in the sonar's own noise level.

TSM-2021 or IBIS III, DUBM-21B can operate at about 6 kts speed for search. It sweeps 360 degrees in sectors of 30, 60 and 90 degrees. Beam width is 1.5 degrees. Range is about 100 to 600 meters from the ship bow. Accuracy in vertical is less than 12 minutes of arc for roll and pitch of ±15 degrees roll and ±5 degrees of pitch. The classification sonar operates at 400 kHz and has a beam width of 0.17 degrees.

The use of non-reflective coatings on mines is rendering echo-classification techniques somewhat ineffective. But such coatings have no effect on the acoustic shadow cast by the mine, and it is said that the extremely precise angular definition of the Thomson CSF Co. sonars TSM series (within 0.17 degrees) permits the mine to be classified from the acoustic shadow.

DUBM-40B minehunting sonar. An all round, 360 degree or variable angle sector scan sonar. Normally is used to sweep a sector of 100 degrees ahead of the vessel, which proceeds at a slow speed. The detection range of the sonar in active mode is more than 100 meters for a two meter circumference metal target. Can use shadow-type classification of targets. Has a small display which shows a magnified view of the acoustic echo return. Also has a PPI (Plan Position Indicator) scan detection presentation.

DUBM-41 & 42 Minehunting sonars. (See Fig. 4.5.) Builder is Thomson-CSF. These are side-looking sonars. Three towed bodies called "Otters" are used. The DUBM-42 is an improved version of the DUBM-21B sonar. It utilizes 3 sonars, side and front looking. The towed submerged vehicles are maintained at constant height above the sea bottom.

An INS (Inertial Navigation System) is integrated into Syledis System and Navstar receiver, autopilot and improved PAP-104 remote vehicle. The improved PAP-104 has a sonar, TV camera, cutter and explosive charge for use in localizing the mines and disposing of them. The Mark V is the latest version of the PAP-104. The later sonars are planned for the BAMO (Batiment Anti-Mine Oceanique) GRP Catamaran-type mine vessels. These will be oceangoing vessels.

A derivative of these French Navy sonars will be used in the new U.S. Navy MCM Avenger Class minehunters. Thomson-CSF is working with the U.S.A. company Raytheon on the SQQ-32 sonar. Raytheon will build the detection sonar, and Thomson-CSF will build the classification sonar. In operation any minelike object will appear, distinctively marked, on the console screen of the computer-aided detection sonar. Then the classification sonar provides a computer-processed, high-resolution picture of that object. May be using parametrics in some circuit designs for this sonar.

Germany

DSQS-11H built by Krupp Atlas Elektronik Co. Used on Federal German Navy type 343 minehunters. RAN (Royal Australian Navy) has also purchased for use on their "Bay" class Catamaran minehunters in conjunction with the MWS-80 portable minehunting shelter. The sonar is hull mounted and can be extended below the vessel. (See Fig. 4.6.) The Royal Thailand Navy has also procured.

MWS-80 Atlas Elektronik Co. Modular Minehunting Sonar System. Ideal for COOP (Craft of Opportunity) installations, as the system is modular and can be moved rapidly from one vessel to another or can be installed quickly on a vessel. (See Figs. 4.6 and 4.7.)

AIS-11 sonar system built by Krupp Atlas Elektronik of West Germany features an "Aladin" system with phase measurement circuitry that provides better definition of underwater mines. (See Figs. 4.8 through 4.11.)

This is a very-high-frequency sonar installed typically in the forward section of the PAP-104 series of remote minehunting vehicles produced in France. Because it is a modular assembly, the AIS-11 sonar can be installed in any vehicle where the mission requires it and space for installation is available, with needed power, etc.

For the incorporation of the AIS-11-104 Active Identification Sonar into the PAP-104 Mine Disposal System, the ship-to-vehicle communication and the power supply of the PAP-104 are modified. The altered PAP-104 can be used either with the sonar bow section or with the television bow section. The AIS-11 uses miniaturized electronics. On-board ship equipment consists of the test and display unit, and an operator's unit, which are adapted to the main control unit of the PAP-104 system.

When it is decided to launch the PAP-104, the AIS-11-104 sonar is checked on board before deploying the vehicle. Simulated sonar signals are injected into the receiving array by means of a special test transducer. If all the electronics work properly, a defined pattern appears on the display unit. The PAP is steered by guidance signals from the main control console. The vehicle heading is also displayed digitally on the AIS-11 display unit. Various sonar range settings are available. In the final phase of classification and identification, the range setting is changed to provide maximum resolution. For precise control of the main disposal charge release, special marks are displayed.

East Germany

For minehunting vessels of East Germany and other East Block countries such as Poland etc., Soviet sonars are used for the most part.

Great Britain

Type 193M VDS (Variable Depth Sonar). The most common British sonar for minehunting. The builder is Plessey Marine. The search frequency is 100 kHz; the classification sonar uses a frequency of 300 kHz. (See Figs. 4.12 and 4.13.)

Type PMS-75 Speedscan system by Plessey Marine. Used in Falklands/Malvinas conflict in the South Atlantic. Enables sector scan sonars to operate in a sidescan mode.

Type 2093 minehunting sonar. Built by Plessey Marine Ltd. This is a dual frequency sonar. Either hull mounted or variable depth. Used on Royal Navy "Sandown" class SRMH (Single Role Minehunter). Designed to operate in deep water and against a full range of sea bed and environmental conditions. The use of advanced sonar technology, integrated with the ship's command system, permits simultaneous search, detection and classification operations while its ability to cover a wide area will allow an improvement in current route survey operations. (See Fig. 4.14.)

Plessey Marine Ltd. collaborated with the General Electric Co. of the U.S.A. in the development of the 2093 sonar.

Type 3010 mine detection sonar. A sidescan type. The builder is Waverly Division of Dowty Maritime Systems Ltd. The 3010 is an update of the type 3000 and is configured for minehunting and route surveillance. This sonar has been integrated with a Quibit TRAC IVB navigation processor and has been used in some trials. (See Fig. 4.15.)

Following a U.S. government competition to select a sidescan sonar and tracking system for U.S. Navy's Craft of Opportunity (COOP) and U.S. mine classification and route surveillance and mapping programs, Dowty Maritime Systems' Waverly Division was awarded a contract for seven of the 3010/T systems. The Dowty 3010/T system is based on the 3010 MCM sonar, but includes additional features such as the Ferranti Trackpoint II towfish tracking system and the latest developments in operator interface and sonar range enhancement. The system uses an image-enhanced color video display with target marking.

The COOP concept is aimed at providing an economic, rapidly deployable mine countermeasures capability, particularly for those navies lacking a fleet of dedicated minehunting or minefield clearance vessels. The equipment is relatively lightweight and can be installed on any small craft that has an insignificant magnetic signature in comparison with most warships.

MSDS—Marconi Space and Defense Systems Ltd. has a 360-degree scan minehunting sonar. Polarized plastic transducers are used that operate from 1 to 1000 kHz. This is a dual-frequency sonar. The search sonar operates on 83.3 kHz. and has a range of 1.2 km. The azimuth resolution is 3.3 degrees. The classification sonar operates at 250 kHz, has a range of 150 meters. Resolution is to 1.1 degree.

The acoustic projector and circular array have 100 plastic receiving elements. Switching is done electronically for beam forming in rapid repeated sweeps through 360 degrees of azimuth. This takes place in a single burst of ultrasonic energy or ping. The sonar transmitter and receiver contain no moving parts.

The sonar can be hull mounted, but MSDS proposes towed arrays. Detection is then enhanced by the larger shadow created by the 15-degree elevation beam at oblique angles. The towed array is also decoupled from the towed vessel's own noise, so gain is improved. The towed assembly is 1.2 meters long and 0.32 meter in diameter. Weight is 90 kg.

A CRT type display is used and spirally scanned in synchronization with the receiver element array. A ring of light appears to spread from the display center for the PPI display. Digital electronic processing is utilized.

RASAT—a programmable universal sonar target for alignment checks on hull-mounted and dipping sonars. The Royal Navy plans to purchase 22 such units.

Italy

Type P-2072 variable depth sonar. Built under license from General Electric Co. in U.S.A. Builder is FIAR Italy Spa. An improved version of the General Electric SQQ-14A. Uses 40% of the later SQQ-30 improvements by General Electric Co.

Japan

Older minesweepers obtained from U.S.A had U.S. Navy type sonars such as UQS-1, SQQ-11, etc. Tendency of JMSDF (Japanese Maritime Self-Defense Force) is to buy a few of a particular sonar, then build (under license) their version of it, adding a few improvements of their own.

Because Japan is actively pursuing new designs of minehunters, it can be expected that they will undertake new sonar designs of their own, possibly incorporating some of the European and U.S.A. sonar techniques and circuits.

Korea (South)

The Swallow shipyard in South Korea is building a catamaran-type-hull minehunting vessel. Prototypes of this design have already utilized the British 193 sonar built by Plessey Marine Ltd.

A number of these vessels may be built both for own-country use and for export. The various sonar minehunting systems are being considered for the production run of this vessel. Included among them is the Krupp Atlas Elektronik DQS-11H with the MWS-80 Mine Warfare Weapon System.

Sweden

On the Royal Swedish Navy "Landsort" class of mine-countermeasure vessel, the French Thomson CSF 2022 Sonar is used. The Royal Swedish navy has ordered MAINS (Minehunting Action Information and Navigation Systems) from Racal-SMS Ltd. of Britain. MAINS forms the heart of the Philips Elektronikindustrier AB (PEAB) Co. 9MJ400 mine-countermeasures system. PEAB is a Swedish company.

The MAINS, when used in conjunction with a T-CSF 2022 sonar as part of an advanced integrated minehunting system with full navigational ship control, command displays and automatic recording capability forms perhaps the most modern planned to date by any navy.

Thailand

The Royal Thailand Navy has purchased some of the Model DQS-11H German Sonar with the MWS-80 Mine Warfare System to be used on their minehunting vessels.

U.S.A.

UOL Underwater Object Location Program for U.S. Navy by General Electric Co. 1942–1950.

UQS-1 Mine Detector. Vintage 1950-1958. Built by General Electric Co. Some could still be in use on older minehunting vessels, particularly those sold or given to foreign navies. (See Fig. 4.17.)

The UQS-1 and UQS-1A are high-resolution scanning sonar equipment used to detect and locate underwater mines for the purpose of avoiding, sweeping or recovering them. The operator can resolve detail in the azimuth direction to two degrees, and in range to 2 meters. In vertical depression, accuracy to 10 degrees. Range is read on a counter after positioning a cursor line by a handwheel. Range circles on the cathode ray scope give approximate range to target. Relative bearings are read from a bearing ring around the viewing tube.

The underwater sonar transducer assembly can be moved in bearing 360 degrees automatically, 90 degrees automatically, or trained manually. The assembly tilts from −5 to +50 degrees in depression, and the power output of the transmitter can be varied. This was a vacuum-tube-type electronics system.

SQQ-11 An older minehunting sonar.

SQS-15 VDS (Variable Depth Sonar). Vintage 1952-1955. Was manufactured by General Electric Co.

CXRP Mine classification sonar. Vintage 1956-1958. Limited use. Manufacturer was General Electric Co.

SQQ-14 Minehunting sonar. A variable depth, dual-frequency sonar for detecting and classifying bottom mines in shallow water. It uses a towed body in the shape of an elongated sphere towed through the center well of a U.S. Navy minesweeper. Manufactured by General Electric Co. Also used by many foreign countries, either built under license or supplied in an older mine vessel that was transferred to the foreign country by the U.S. Navy. (See Figs. 4.18 and 4.19).

The towing cable of this sonar is unique. It consists of half-meter sections of articulated struts with universal joints at each section, permitting the cable to flex in any vertical plane, but preventing movement from torsional motion. This gives a constant heading to the towed body and eliminates the need for a gyroscopic heading reference system. A rubber-coated electric cable containing 35 shielded, coaxial, and individual conductors passes through the center of the strut sections, terminating in a slip ring in the towed body, and in a winch with compound winding drum in lieu of a slip ring installed on the operating deck of a mine-countermeasures vessel.

The articulated strut type of tow cable has built-in problems relating to winching, storage on the winch drum, and the ship's towpoint-type construction. There are 117 struts in the system. These are stored in a single layer on a cylindrical drum equipped with specially designed pads serving as contact points for the articulated strut knuckles. Since a conventional level-wind mechanism cannot be used with the struts, the winch drum itself is designed to traverse axially by means of an Acme lead screw and rotating nut, thereby accomplishing the level and wind functions and maintaining a fixed feeding point. All components of this system must be as nonmagnetic as possible. Struts, bearings and knuckles are fabricated of high-strength Inconel alloy. The winch drum is cast aluminum and the supporting structure is aluminum and stainless steel.

Approximately 70 systems have been manufactured under U.S. Navy contracts. The SSQ-14 sonar systems are produced under license in Italy by CGE Spa, and supplied to the Italian Navy. The Italian Navy has also modified these sonars to incorporate some of the later improvements in sonars such as the SQQ-30. Other countries that have these sonars installed are Spain and Belgium.

SQQ-30 Minehunting sonar. Follow-on to the SQQ-14 General Electric Co. system. (See Fig. 4.20.) Uses search and classification sonars on different frequencies; search is lower frequency, classification is higher. The towed body is a hydrodynamic egg-shaped vehicle. It is stored in a well in the forward part of the vessel. Two displays are utilized; one for detection, the other for classification.

The MCM-1 Avenger class vessel now entering the U.S. Navy will have the SQQ-30 installed on the initial few. Follow-on will be the SQQ-32 sonar.

The SQQ-30 sensor is limited to speeds in the 2 to 5 knot speed range. Recent tests demonstrated a 50% probability of detecting a mine in a single pass.

SQQ-32 AMSS (Advanced Minehunting Sonar System). The manufacturer of the search sonar is Raytheon Co. Thomson-CSF of France builds the classification sonar. (See Figs. 4.21 and 4.22.)

This system is under technical evaluation now. First Fleet installation may be 1990-1991. To be installed on minesweepers of MCM-1 Avenger class and possibly later MHC-51 "Osprey" class which is a derivative of the Italian Navy "Lerici" class.

Two sonars are used in the SQQ-32. The search sonar uses a lower frequency, and the classification sonar frequency is higher for better resolution. Beam forming, signal processing and modular packaging of the electronics is utilized.

The Raytheon search sonar is used to search for mines over a wide range. Has computer-aided detection capability. Can discard most non-mine targets. Computer interfaces to display consoles and does system integration. The Thomson-CSF-Sintra classification sonar is derived from the DUBM-21 or TSM-2022 minehunting sonars. The high-frequency transmission gives high-quality definition of the targets, as seen on the displays.

The wet end of the SQQ-32 sonar is in a hydrodynamically shaped vehicle. It is deployed on a tow cable. The winch is provided by Charles Stark Draper Labs of Cambridge, MA in the U.S.A. They also design the deployment retrieval system, an armored tow cable and winch, as well as fabricate the wet end consisting of the towed vehicle. The greater sensitivity and other equipment improvements enable the SQQ-32 to be towed at higher speed than the SQQ-30 sonar, on the order of 6 to 10 knots.

Raytheon Co. will provide the computer facilities, display consoles, and overall system integration. In operation, a mine-like object on or near the bottom will be distinctively indicated on the console screen of the computer-aided detection sonar. Then the classification sonar provides a computer-processed, high-resolution picture of that object. The SQQ-32 display consoles can be used with either the search or classification sonar.

Mk-24 Underwater Ordnance Locator sonars. Built by Klein Associates of Salem, New Hampshire, U.S.A. Sixteen of these Mk-24 systems have been sold to the U.S. Navy's Explosive Ordnance Disposal facility, as part of the COOP program. (See Figs. 4.23 and 4.24.)

Designed and built for the offshore industry, the Mk-24 is a versatile sidescan sonar which can be deployed from unmodified commercial vessels. It has five basic components:

- Model 521T recorder
- Two-channel transceiver and graphic recorder
- Model 422S-101 100 kHz $\frac{3}{4}$-degree beam-width sidescan sonar towfish
- Model 604 alphanumeric annotator
- K-Wing I hydrodynamic depressor.

The plotting and relocation of sea bed features and mine target is ensured by the use of an ultra-short-baseline towfish tracking system and accurate above-water navigation. Frequencies are on the order of 100 to 500 kHz. The Klein sidescan sonar has also been deployed out of a helicopter of opportunity—such a those used by the U.S. Coast Guard. It has been approved for service use and designated Mk-24 Mod 0 by the U.S. Navy. The Mk-24 system is available off-the-shelf, including spare parts, for under $100,000. It provides superior performance for minehunting as well as channel conditioning, port break-out and support to amphibious assaults.

BURMMS Buried Minehunting System. Exploration of technology contract from Naval Coastal Systems Center (NCSC) at Panama City, Florida with Westinghouse Co. of Baltimore. Intent is to operate from 100 to 200 meters in depth. For detection, classification, localization and marking. Would use a bottom penetration sonar, and a superconducting cryogenic magnetometer. Precise navigation is required.

AQS-14 Airborne sonar, used for minehunting. Carried by helicopters such as RH-53D and MH-53E. The builder is Westinghouse Electric Corp., Defense Group, Oceanic Div., Annapolis, MD. (See Figs. 4.25 through 4.31.) A number of these minehunting sonars are being procured for the U.S. Navy airborne minehunting program. The AQS-14 is a three-component system:

Active stabilized underwater towed vehicle
Electromechanical tow cable
Airborne electronics console.

The system can be installed or removed in less than two hours. The three-meter-long towed fast-search vehicle carries a multibeam side-looking sonar. Electronic beam-forming technology is used to provide focusing for all ranges. The adaptive processor allows hands-off operation through varying bottom reverberation environments.

In towing, the sonar maintains an operator-selected depth above the sea bottom through an active control system and specific sensors. Self-protection circuits prevent the vehicle from broaching or striking the bottom, while a closed-loop roll control adds additional vehicle stability. An on-board dispenser holds acoustic beacons that can be dropped individually on command to mark objects or areas of interest. The armored, non-magnetic tow cable has been designed to have a small diameter and a low drag coefficient by using a full multiplexer. Towing altitude is approximately 75 to 100 feet. Normally operated in daytime with clear weather conditions. Two operators man the control-monitor console in the forward part of the helicopter. The unit includes a television moving-window sonar display, underwater vehicle controls and status displays, system status indicators, magnetic tape recording for sonar and other mission data, tape playback and controls, and the built-in test equipment controls and displays.

Rapid tow speeds and the coverage of large areas result from the sonar's multiple beams, but at lower tow speeds the towed sonar automatically reduces the number of transmitted beams. This beam-former provides focus throughout the sonar range so that sea bottom highlights are sharp and easy to interpret. The adaptive time-varied gain processor gives consistency of display background in spite of wide variations in sea bottom texture and material, with little or no manual adjustment. System functions cover all tasks needed to locate, classify, mark, permanently record and review records of mines, mine-like objects and underwater terrain features.

The AQS-14 system was deployed on board the RH-53D minehunting helicopters aboard the *U.S.S. Shreveport* LPD 12 to the Gulf of Suez in August, 1984. The mine warfare teams and platforms were airlifted by C-5 transports for deployment with the *Shreveport* from Rota, Spain. The system was normally operated in winds of 30 to 35 knots, and in one case was streamed and recovered in winds over 45 knots with 12 to 15 foot swells.

This system could also have applicability to hovercraft, hydrofoils, drones, and other operating platforms. The ALQ-14 was developed and built for the Airborne Mine Defense Program Division of the Naval Air Systems Command.

AQS-17 This is a relocation sonar. An active sonar with receiver. The range is approximately 1200 meters. Operates at a frequency of 200 kHz. Operating depth is from 3 to 50 meters. The main body is a 1-meter-diameter sphere, weighing 300 kg. in air, 100 kg. in water. The consoles for operation are about 100 kg. in weight. Could tow in some C-4 explosive to destroy the target mine. Note: A conceptual system.

PAMS Pressure Acoustic Monitoring System. This system uses an "A" size sonobuoy, either surface or aircraft deployed, to monitor acoustic signals from a minesweeping projector, pressure variations caused by sea wave motion. A novel system to aid in minehunting operations.

PQS-2A A continuous-transmission, frequency-modulated sonar designed to be portable and hand held by explosive ordnance disposal divers to locate and close on mines and other objects.

The system was designed by Naval Coastal Systems Center and is produced by General Instrument Co. The unit weighs 4 kg. and can operate in depths to 100 meters. Other specifications are as follows:

Range:	
Active mode	20, 60 or 120 meters
Passive mode	2,000 meters
Transmit frequency	115 to 145 kHz (CTFM)
Passive detection	Detects 39 kHz signal source with a level of 150 dB/μpa at 1 meter, at a minimum range of 1,828 meters
Magnetic signature	MIL-M-19595
Output technique	Tone or pulse audio in earphones, frequency variable with range
Power source	Rechargeable batteries
Operating depth	Up to 100 meters
Operating temperature	−2°C to +30°C
Weight:	
In air	4 kg.
In water	Slight negative buoyancy

WQS-1 Ahead-look sonar used on the SLQ-48 MNS (Mine Neutralization System) underwater remote-controlled minehunting system. The designation of WQS-1 refers to a family of small size, medium resolution, multibeam, cylindrical array sonars developed by ARL, University of Texas in Austin, for use as obstacle avoidance sonars in remote-controlled underwater vehicles. The original WQS-1, developed in 1970, operated in depths to 60 meters; deeper-water versions were later developed.

The display on the vessel from which the MNS deploys uses a scan converter and RAM (Random Access Memory) to refresh the image as opposed to a phosphor for image persistence.

The MNS has cutter arms in the bow. At tilt angles below −30 degrees, the cutter arms interfere with the projector and receiver beams.

General specifications for the MNS WQS-1 are as follows:

WQS-1 MNS Specifications

Type: Multibeam, single frequency, short pulse, within-pulse beam scanning.
Frequency: 200 kHz
Pulse Length: 0.25 ms.
Receiver Bandwidth: 10 kHz
Noise Figure: 12 dB
Minimum Detectable Signal: +55° dB re 1 μPa at hydrophone face with 0 dB signal-to-noise ratio.
Range Resolution: 19 cm.
Range Scales: 3, 10, 30, 100, and 300 yards
Tilt Angle: Variable in 5° steps from +90° to −60°
Source Level: Variable in 31 steps to 200 dB re 1 μPa at 1 m. maximum
Projector Beam Pattern: 18° vertical by 120° horizontal at −3 dB points
Receiver Beam-former: Delay lines form 40 beams on 3° centers, covering a 120° field of view
Receiver Beam Patterns: 18° vertical by 4° horizontal at −3 dB points. Uses Hamming shading in the horizontal plane for theoretical side lobes suppression of 42 dB. Has a grating lobe response at ±50° about 15 dB below the main lobe. Far field is 2 meters.
Display: RAM-refreshed display in operator console. Sonar image displayed in a modified PPI wedge shape and also as a gunsight view. Annotations include range and azimuth markings, tilt angle, transmit level, water depth and vehicle altitude.
Power Requirements: 3 amps. max. at 19 to 29 Vdc.
Physical: Cylinder, 35 cm. diameter, 20 kg. in air, 4 kg. in water.

U.S.S.R.

The first of the minehunting sonars used by the Soviets were lend-lease during World War II.

TAMIR-11. A Soviet Navy high-frequency sonar. Vintage 1950. Used in most Soviet-built minehunting vessels that had sonars after World War II.

Present-day minehunting sonars. High frequency, lightweight, operate on 60 to 500 kHz frequency range. Short range, high definition. Ranges vary from 100 to 250 meters, at depths of 10 to 60 meters.

Airborne minehunting sonars. It is believed that some of the Soviet Naval helicopters can be and have been equipped to deploy towed sonars similar to the AQS-14 system of the U.S Navy. Their ships have deployed with helicopters aboard to participate in mine-clearing operations in various locations in the Middle East. They certainly are doing more than simply visually looking for mines below the surface or on the surface. The Soviet Union has several helicopters of medium and heavy lift capability that could tow sonars.

Figure 4.1. The display panel of the French Thomson-CSF DUBM-20A minehunting Sonar.

Figure 4.2. The French Navy Naviplot Tactical Table as used with the Thomson-CSF minehunting system. It includes a detection and classification sonar, part of the IBIS V system.

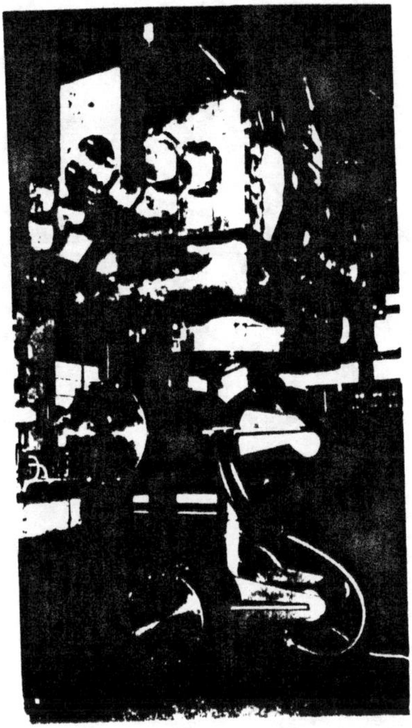

Figure 4.3. The French Navy DUBM-21B double-hull sonar as used in the 40 Tripartite minehunters.

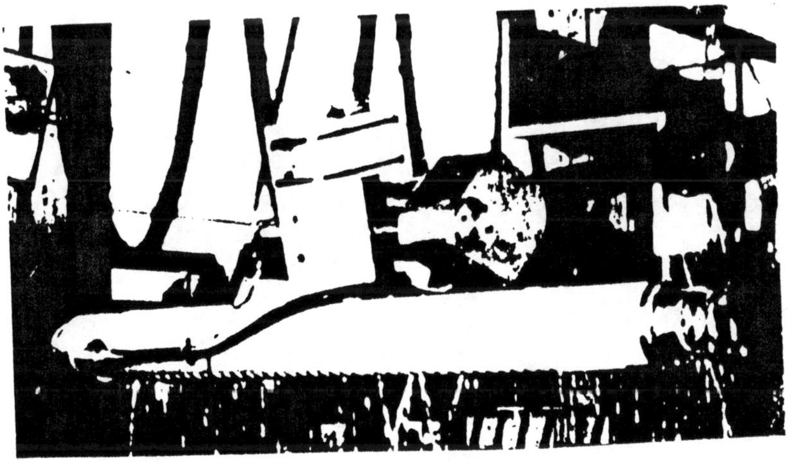

Figure 4.4. The acoustic array of the French Navy Thomson-CSF TSM 2022 minehunting sonar.

Figure 4.5. The French Navy Thomson-CSF DUBM-41B towed body on the *Betelgeuse* MCMV prior to sea trials.

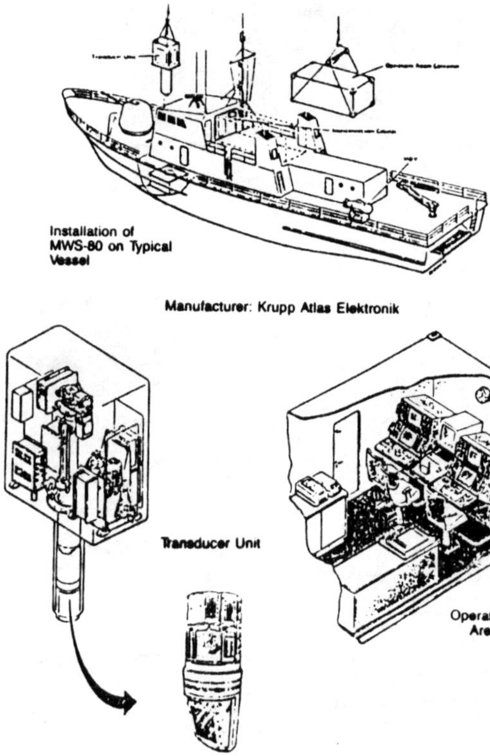

Figure 4.6. New German modular mine-countermeasure weapon system MWS-80. Used with DQS-11H sonar.

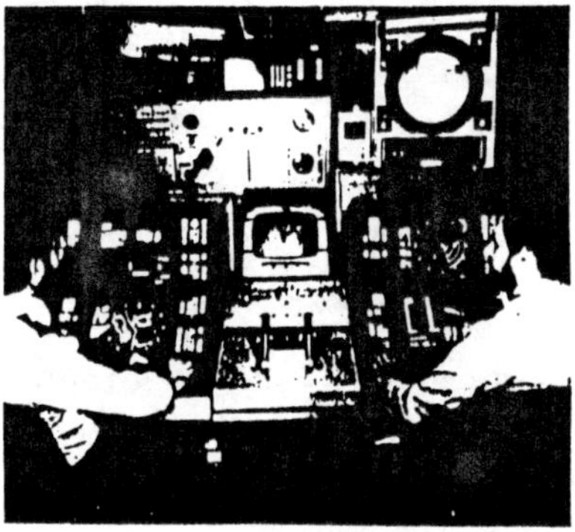

Figure 4.7. Krupp-Atlas Elektronik MWS-80 Mine Warfare Weapon System Module is installed on the RAN "Bay" class MHCAT (Catamaran), and Royal Thai Navy vessels.

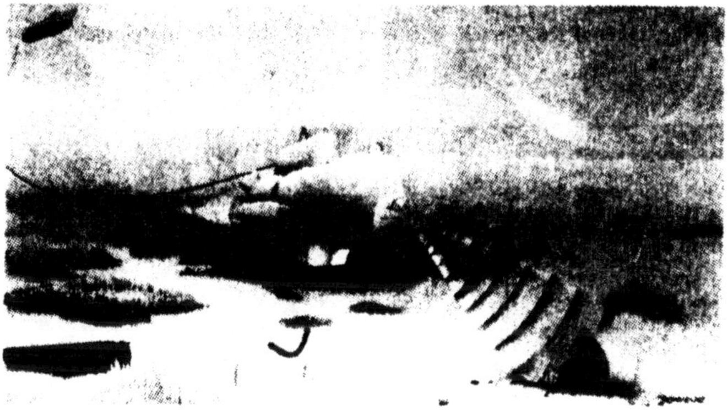

Figure 4.8. The PAP-104 mine disposal vehicle searching for mines with Atlas AIS-11 Active Identification Sonar.

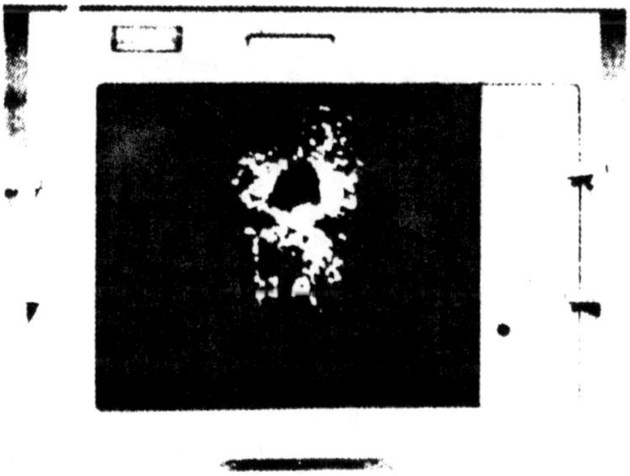

Figure 4.9. The Krupp Atlas Elektronik AIS-11 sonar display.

Figure 4.10. Front end view of PAP-104 showing AIS-11 installation.

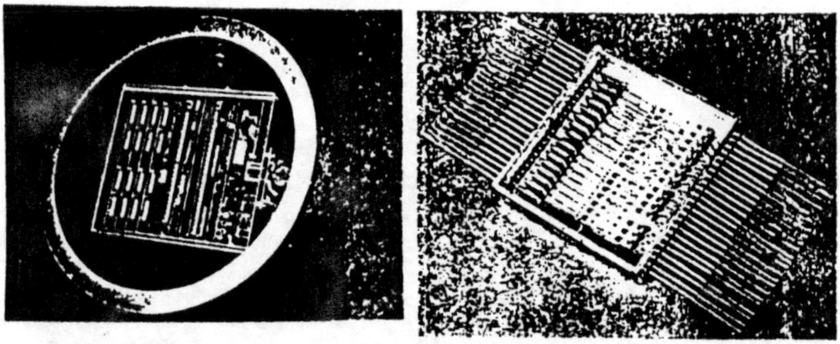

Figure 4.11. Electronics of the AIS-11 Atlas Active Identification Sonar as used in the PAP-104 mine disposal system.

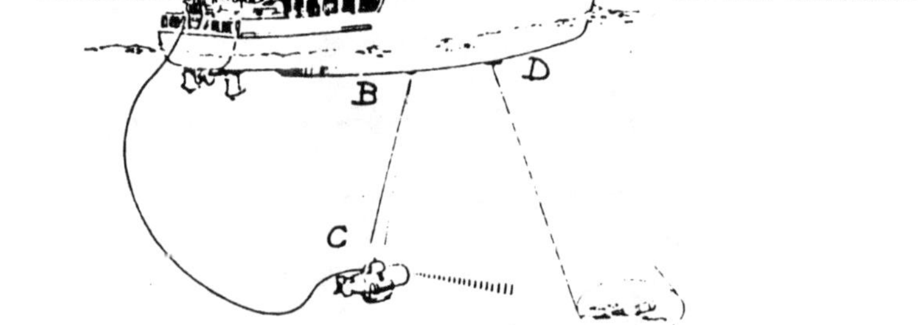

Figure 4.12. A typical minehunter with Racal Hyper-Fix positioning system (A), an Aqua-Fix Suite (B), PAP-104 vehicle (C) and a Type 193M sonar (D).

Figure 4.13. British Plessey Ltd. minehunting sonar Type 193M in operation.

Figure 4.14. A prototype installation of the British Plessey Marine Type 2093 minehunting sonar display.

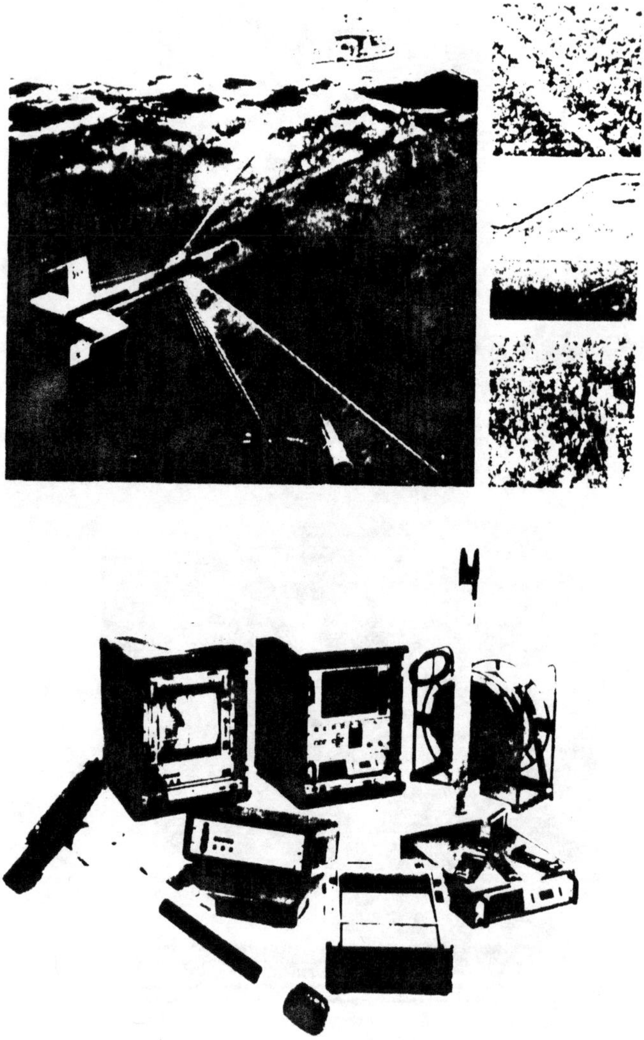

Figure 4.15. British sidescan sonar for mine countermeasurers. Manufacturer is Dowty Maritime Systems Ltd., Waverly Division. This is the Type 3010 sonar.

Figure 4.16. FIAR Co. of Italy minehunting sonar. A license-built version of General Electric Co. of U.S.A. SSQ-14 variable-depth sonar, sometimes called "Squeaky Fourteen."

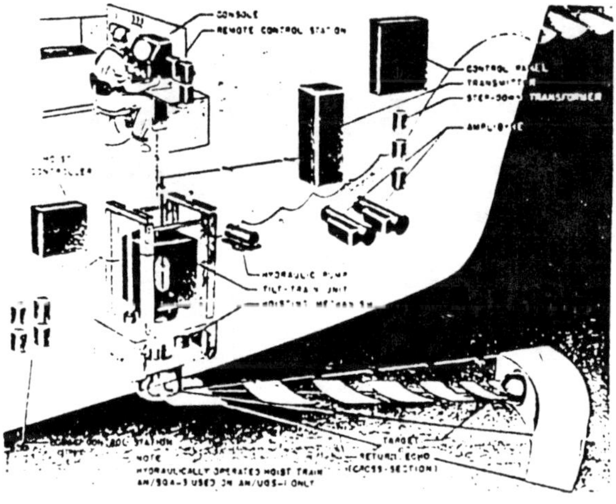

Figure 4.17. Components of the UQS-1 minehunting sonar system.

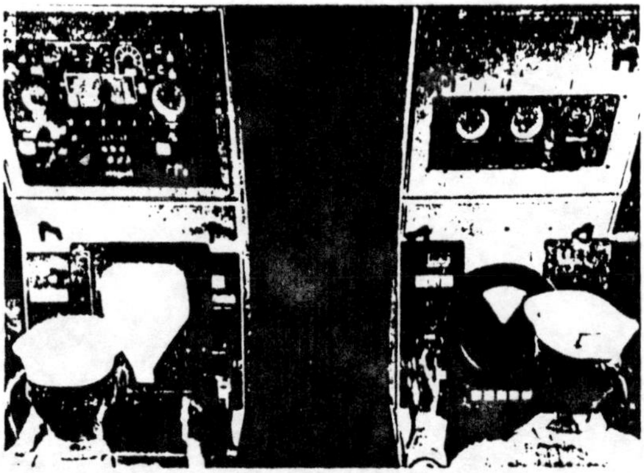

Figure 4.18. The detection and classification consoles of the U.S. Navy SQQ-14 minehunting system.

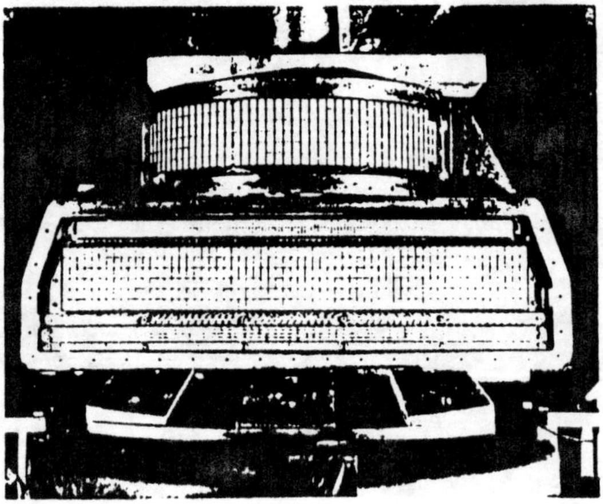

Figure 4.19. SQQ-14 projectors and hydrophone arrays.

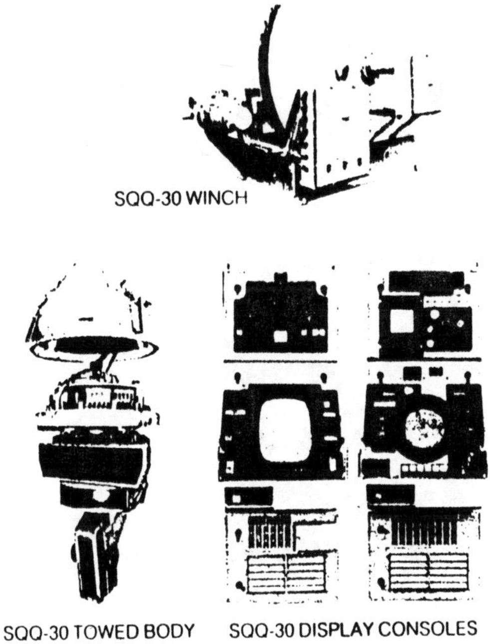

Figure 4.20. Components of the U.S. Navy SQQ-30 minehunting sonar system.

Figure 4.21. SSQ-32 display consoles. Can be used for either search or classification sonar.

Figure 4.22. Wet end of the SQQ-32 minehunting sonar. Planned installation in the U.S. Navy "Avenger" class MCM-1 vessels.

Figure 4.23. The Mk-24 Klein Co. sidescan sonar deployed from a helicopter of opportunity, and deployed from a small boat. The display is of a U.S. Navy Mk-6 moored mine.

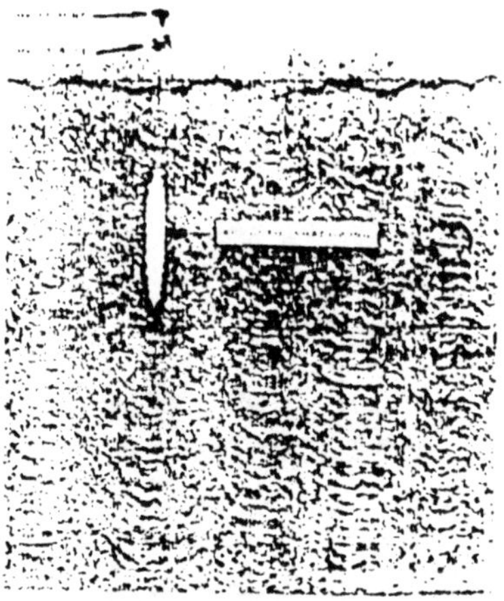

Figure 4.24. The Mk-24 Klein Co. 500 kHz sidescan sonar display showing an Mk-18 base with an Mk-17 mine moored 2 meters above the bottom.

Figure 4.25. MH-53E U.S. Navy helicopter towing AQS-14 sonar.

Figure 4.26. Operator at console of AQS-14 airborne sonar.

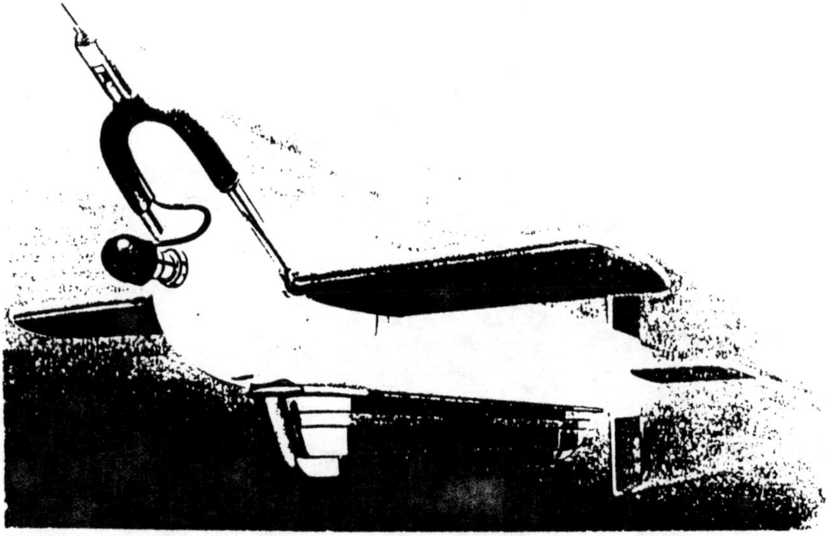

Figure 4.27. U.S. Navy AQS-14 airborne minehunting sonar. Manufacturer is Westinghouse Electric Corp.

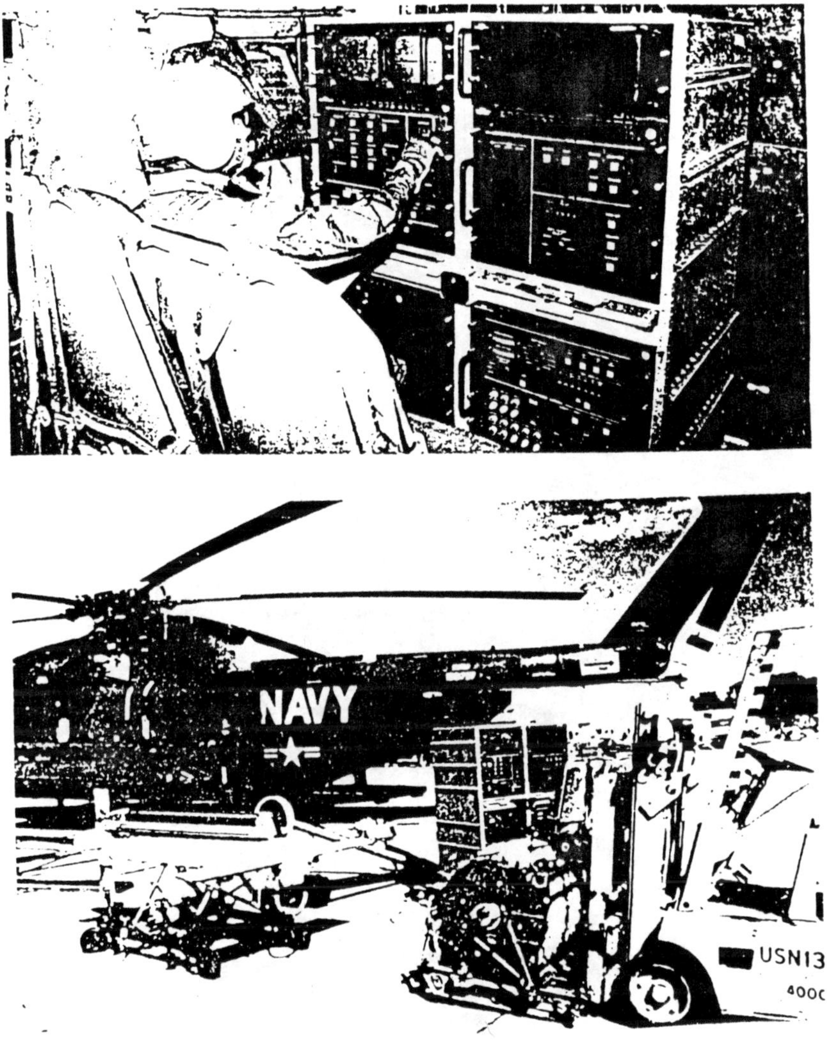

Figure 1.28. Top view shows AQS-14 airborne sonar operator in RH-53D U.S. Navy helicopter at the console. Bottom view shows some of the components of the AQS-14 minehunting system. Sonar probe is at left, and the control/monitor console is seen behind the towing cable assembly on the forklift.

Figure 4.29. A U.S. Navy helicopter towing the AQS-14 sonar. Note the variable depth sweep and resolution scanning areas.

Figure 4.30. AQS-14 airborne sonar display in helicopter, revealing a sunken ship. Note the detail.

Figure 4.31. A section of pipeline as viewed on AQS-14 sonar in Gulf of Suez.

CHAPTER 5

Remote Mine Neutralization Systems

Every ship is a minehunter—once. This statement has been proven a number of times recently in the Persian Gulf. When the ship is a supertanker such as the *Bridgeton*, it can be a minehunter perhaps twice. This was demonstrated for all the world to see by the *Bridgeton* leading the way for its U.S. Navy escorts, even after it had encountered one mine that damaged it. However, supertankers are expensive minehunters.

Undersea vehicles are listed in several categories. They can be manned, or unmanned. They also can be tethered or untethered (autonomous) vehicles in the case of unmanned vehicles. Countering mines is the only large-scale application for military Remotely Operated Vehicles (ROVs). All existing countermine vehicles can place explosives near mines to bomb them. These bomber vehicles are also equipped with cutters for moored mines and sometimes are equipped with manipulators which can be used for recovery purposes.

These underwater vehicles that are unmanned have a distinct advantage over manned minehunting craft. While divers can physically work around mines, this is a very dangerous profession. The tethered or untethered craft can be used in the same way. With television, high-definition sonars, cutters, manipulators, lights and bombs, the task is much safer for humans who can control minehunting and mine destruction remotely.

The French Navy was probably the pioneer in the use of remote unmanned craft for minehunting. The PAP-104 has proven to be a very practical and reliable system. Many foreign navies have adapted it for use from their vessels. The PAP is sailor friendly. It is rapidly and easily deployed and can be operated and recovered with ease. It is a bomber that kills as opposed to neutralizing mines, placing an explosive very near a mine and causing a sympathetic detonation.

In the Mine Countermeasure (MCM) application, there are a number of opportunities for the ROV, including variations as front-swimming minehunting sonars and expendable destruction vehicles. Autonomous ROV (A/ROV) mineclearance concepts from Craft of Opportunity (COOP) could gain in popularity as technology marches on. In 1970, the Royal Navy conducted tests with autonomous vehicles that provided some innovative operating methods and hardware. The idea was for the A/ROV to conduct a side-scan bottom search of a predetermined area and on completion, to surface and radio data back to the mother ship for processing. Sometimes these remote vehicles are called Autonomous Underwater Vehicles (AUVs).

The military sector was the first to commit significant funds and effort to ROVs. In the early 1960s, the Naval Ocean Systems Center (then the Naval Electronics Laboratory) in San Diego, California sponsored construction of the CURV, a tethered, free-swimming vehicle. It was launched in 1965, and the following year it was instrumental in retrieving a lost U.S. hydrogen bomb in about 600 meters of water off the coast of Spain.

The first commercial ROVs appeared about 1974. One of these was the Hydro Products RCV-125, with the later-designated RCV-225 type. There was a boom in ROV production and application in the offshore oil and gas market. However, there was not much action in the military applications area.

Some of the commercial manufacturers entered into the ROV field with development of military vehicles. Two of these were Hydro Products and Ametek/Straza Division. Military applications of ROVs in the 1970s tended to closely parallel those of the manned submersibles. ROVs were utilized for inspection of hardware, sunken vessels and downed aircraft; survey of sites and routes; search/identification/retrieval and retrieval assistance. There was not much U.S. Navy interest yet, but foreign navies began fitting ROVs into their mine-neutralization applications.

The various navies do not advertise their ROV capabilities, as do the commercial operators, but the commercial ROV manufacturer must advertise and make his product known. The following are some of the naval ROV operators:

U.S.A.	Thailand
Canada	Nigeria
Great Britain	Israel
France	Egypt
Italy	Belgium
Japan	Netherlands
Sweden	West Germany
South Korea	U.S.S.R.
Norway	Yugoslavia

The Soviet Union has made at least one 1500-meter ROV, and it is safe to assume that it has made others.

Sometimes ROV operations make the headlines. In 1982, a Soviet "Whiskey" class submarine was bottomed for several days in Swedish coastal waters near Karlskrona. There is a Swedish Navy base nearby. The Swedish Navy at that time had at least four Sea Owl ROVs. This gave them the capability to inspect and videotape the details of the Soviet submarine with ease.

The world's navies probably have on the order of 40 or 50 different types of ROVs that can be used for general purposes. A number of low-cost ROVs have been coming on the market in the last few years, and their usage can be expected to increase. Some idea of the scope of the military impact on the tethered, free-swimming ROV count can be gained by considering that, of a total of about 750 ROVs developed, approximately 325 are for military applications.

Mention has already been made of the pioneering efforts of the French Navy in the area of military ROVs. In 1973, the French firm Societe ECA of Meudon began production of a mine-neutralization ROV called PAP-104. U.S.A. ROV manufacturers and their foreign counterparts took little notice of this until about 1979, when ECA announced the sale of some 120 vehicles throughout the world. By 1981, the count had risen to

160 and today is perhaps over 250. It has a combat-proven record in the Falklands/Malvinas combat, where it performed thousands of missions. By 1981, virtually every major ROV manufacturer had entered the field. There was profit to be made, as the per-system cost of the PAP is about $350,000.

Very little has been heard from the military about untethered or autonomous vehicles. About 23 programs in this arena can be identified from the openly published material. Fourteen of these are supported by the military. At present, most untethered ROVs are primarily developmental craft that are used to identify problems in technology that must be overcome before these craft become a practical and effective adjunct to the existing cable-connected craft. Autonomous ROVs have been used by the U.S. Navy since the early 1960s but, as with towed vehicles, their numbers are limited.

More recently, the autonomous ROV has been the subject of greater interest by the military. This had been due to advances in electronics—mainly the microprocessor. Since this type of craft is designed to operate at distances beyond the visible range of the station or vessel from which it operates, it can conduct missions covertly and even under ice. This is of particular interest to the submariner.

Untethered vehicles presently operate in one of two modes. One is pre-programmed, or autonomous. A task is given to the craft's microcomputer, which directs the vehicle in terms of its depth, course, speed, what data it is to collect and when to return. All of these events are accomplished without intervention from the base station or ship. A second operational mode consists of controlling the craft's course, depth and dive duration via an acoustic link to the base station or ship. Most data collected by the ROV is recorded in the vehicle and analyzed when it returns to base or ship. Some applications may require that the data be telemetered to the base or ship in real-time as the mission progresses. This is certainly true if time is of the essence.

The entry of the Low-Cost ROV (LCROV) into the military market could be very important. Costs range range between $25,000 and $50,000, so it may be possible to have an expendable mine-neutralization ROV. As mines become more intelligent, it is conceivable that in the future a vehicle such as an ROV approaching a mine may be blown up by the mine. This may be a practical approach, and cost-effective, as well. Another consideration is that some of the later mine-neutralization vehicles weigh over 1000 kg, a weight that requires a ship of moderate size to accommodate the launch-and-retrieval system needed to handle the vehicle. The result is that the minesweeping vessel must be dedicated to the task of operating one vehicle at a time. The LCROV, which can weigh less than 20 kg and can be powered by standard 12-volt batteries, can be used from small vessels and is air-transportable.

Note: A good summary of ROV craft is contained in *ROV Review 1988—Third edition*. Available from: Mr. Deam Given, Publisher, Windate Enterprises, Inc., PO Box 368, Spring Valley, CA 92077. Tel: (619) 660-0402.

Technology will help solve the mine-neutralization problem. One can cite an example from World War II to prove the point. Figure 5.1 shows a picture of a British "Wellington" bomber that was modified to combat German magnetic mines. It carried a large, 15-meter diameter circular magnetic loop beneath the wings and the fuselage. This was energized with current from an auxiliary generator driven by a motor in the fuselage. As the aircraft flew low – as low as 60 feet above the surface – the powerful magnetic field from the loop set off the mines. There was of course the chance that the Wellingtons could be brought down by the blast, right under their tails. They did,

Figure 5.1. British RAF "Wellington" Bomber aircraft configured with large magnetic loop beneath the wings and fuselage. Used for detonating German magnetic mines during WW II.

however, have considerable success later in the war clearing magnetic mines that German aircraft had dropped in the Suez canal.

The following is a listing of some of the remote mine-neutralization systems being used by the world's navies. The listing is alphabetical by countries.

Australia

Klein 590 system. The Royal Australian Navy has selected this system for their Craft of Opportunity Program. This uses a side-scan sonar towfish vehicle. This is a U.S.A. development.

PAP-104 tethered craft. Has been selected for use on Australian mine-countermeasure vessels. This is a French remote-controlled vehicle.

Belgium

PAP-104 remote-controlled mine-disposal craft. Belgium uses the PAP-104 manufactured by the French company ECA on their "Tripartite" minehunting vessels.

Canada

Trailblazer remote-operated vehicle. Constructed by Submarine International Ltd. of Canada and associated Farey Systems of Great Britain and Underwater Systems of Australia.

Finland

Fiskars, Espoo (near Helsinki) builds an integrated minesweeping system, consisting of a magnetic and an acoustic sweep. Provides strong magnetic flux without risking safety of the towing vessel. The acoustic sweep covers a broadband frequency at high acoustic pressure. A microprocessor controls the sweeping current magnetics and the acoustic equipment. The magnetic cable assembly is about 500 meters and is normally towed at a speed of about 7 knots. The acoustic assembly is 440 meters long.

France

PAP-104 (See Figures 5.2 through 5.5) PAP = Poisson Auto-Propulse. With over 280 units ordered by 10 navies and over 5000 successful runs, the French Navy PAP-104

Figure 5.2. French Navy Remote-Control Submersible PAP-104. Built by ECA Co. for mine countermeasures.

has set a standard in mine-neutralization vehicles. The builder is ECA Co. A wire-guided, submerged vehicle, self-propelled. Equipped with a searchlight, television camera, an explosive cutter for moored mines, and a 100-kg disposal charge to blow up bottom mines.

The PAP-104 carries a destruction explosive charge and weighs, with the charge, 700 kg and is 2.7 meters long. Operating range is 500 meters at depths of 100 meters or less. The maximum speed is 5.5 knots. The depth of the vehicle is controlled by a drag rope which stretches down to the bottom and maintains the vehicle at a set depth. The vehicle is guided by the parent minehunting ship to the position of a mine, where the searchlight and TV camera or sonar send a picture back to the mother ship giving details on the mine-like object. If it is determined that the object is a mine or if it is suspected to be one, the explosive charge is dropped as nearly as possible to it by remote control from the ship. The PAP-104 is then guided back to the ship or to some safe distance, and the charge is exploded, again by remote control, and should blow up the mine by sympathetic detonation. A Rheinmettal Co. remote-controlled fuse is used to detonate the PAP-104 demolition charge. (See Figure 5.5 for this device.)

The Mark V is the later version of the PAP-104. It has computerized navigation aids to assist the pilot (ship operator), a fiber optics bobbin which can carry up to 4,000 meters of cable, redesigned electronic control unit and teletransmission system, two cutters, one on either side of the vehicle, a new tilt Low Light Level Television (LLTV) camera, which can be used simultaneously with a color TV camera, automatic heading, special device to facilitate recovery, positioning device and a depth indicator and radio control that allows the PAP vehicle to be brought back to the ship in case of cable failure. Two vertical thrusters are now used. Three different heads (standard, short-range sonar, relocation sonar) can be utilized. The operational depth has been increased to 300 meters. Built-in test equipment is provided.

Figure 5.3. A French Navy MSO in the Persian Gulf is shown launching the PAP-104 wire-guided submerged vehicle. The weapon underneath is a 100-kg disposable charge to blow up bottom mines. Note the simple handling system.

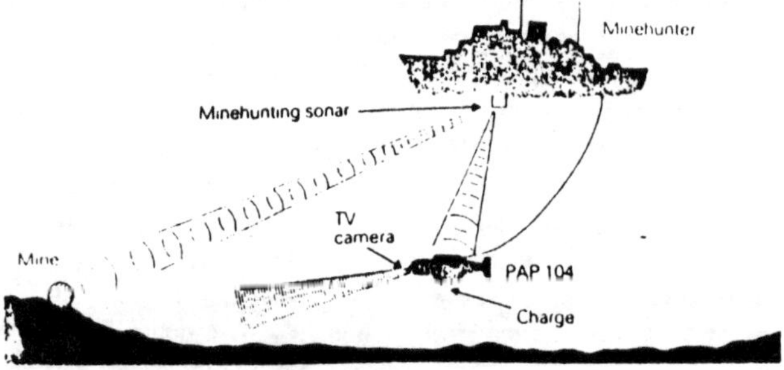

Figure 5.4. The French Navy PAP-104 System. First the ship's sonar detects a mine-like object. The submersible is guided to the object using remote control. A demolition charge is then placed beside the mine and detonated.

Figure 5.5. A Remote-Controlled Fuse is used to detonate the PAP-104's demolition charge. Rheinmetall makes the fuse.

Germany

Pinguin B-3 remote-controlled wire-guided unmanned submersible. (See Figure 5.6.) This is a joint company effort with VFW Fokker and MBB (Messerschmitt-Bolkow-Blohm GmbH), both of West Germany. It carries two 100-kg explosive charges

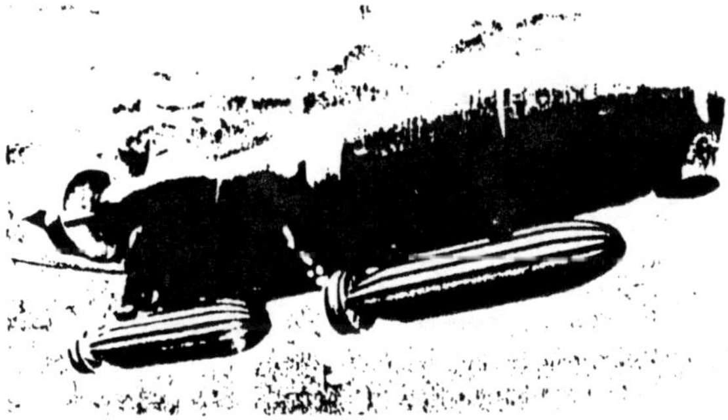

Figure 5.6. The West German Navy "Pinguin" B3 Mine-Disposal Submersible in operation. Note the two disposal charges underneath.

and the guidance cable floats on the water to avoid snagging the sea bottom. The vehicle is guided from information obtained by the ship's sonar, guidance being either automatic or manual, and as in the PAP-104, Pinguin is positioned over the mine, where one of the charges is dropped. The charge apparently explodes by means of a delay fuse. After one charge is dropped and one mine is disposed of, it can go on to dispose of another. The vehicle is self-propelled by two 3-kW units and it also has a controlled 0.5-kW lift engine. The power supply is by a silver-zinc battery housed onboard the vehicle. To assist sonar tracking, the Pinguin is equipped with a sonar transponder and it also has two searchlights and a TV camera. Maximum diving depth is 100 meters and cruising speed is 6 knots, with a mission radius of 600 meters. The vehicle is 3.5 meters long, weight is 1350 kg. The advanced-type battery provides high performance permitting several long-endurance missions between recharging cycles.

PAP-104 remote-controlled vehicle is also used by the West Germans on their minehunting vessels.

Troika. Remote-controlled small boats guided from a mother mine ship. (See Figure 5.7.) The builder is Krupp MAK Co. of Kiel, West Germany. This system is classed as a simulation minesweeping system. It can sweep silted-up ground mines with acoustic-magnetic or magnetic methods. It features safety and sweep performance without endangering the crew. Detects and sweeps moored mines. The control ship and three unmanned simulation sweep craft comprise the Troika system. It has outstanding shock resistance and large action radius of the sweep craft as well as high availability under all weather conditions. The vessels are 95-ton craft. The control ship directs the course and speed of the remote Troika craft and

Figure 5.7. German Navy "Troika" unmanned remote-controlled minesweeping craft. The "mother ship" is in the background.

automatically deploys their sweeps. Six groups (18 minesweepers) and six control ships are planned.

The remotely controlled craft lead the way through a minefield, simulating a large ship's magnetic and acoustic signature to detonate influence mines. Then the manned guidance ship cuts the mine mooring cables in a mechanical sweep. (See Figure 5.8 for such a wire-cutter device.)

Pinguin Al. Was developed for the German Navy by VFW for quite different, classified, deep-water applications. Has only one propulsor and is therefore dependent on hydrodynamic lift, provided by wing-like horizontal fins, for stability.

Great Britain

PAP-104 remote-controlled mine-disposal vehicle. The Royal Navy is another of the many countries that use this French vehicle. The British used their PAP-104 systems extensively during the Falklands/Malvinas war in the South Atlantic to locate Argentine mines.

Minnow remote-controlled vehicle. (See Figure 5.9.) Manufacturer is Marconi Underwater Systems Ltd. This mine-disposal submersible is similar to the French PAP-104 and German Pinguin but is powered by electricity supplied to it down the wire from its mother ship, giving it unlimited endurance. The other systems have onboard power from batteries and hence are limited in time on station. The Minnow has its own sonar and is only 2 meters in length. It also has fore and aft

Figure 5.8. A Mooring-wire cutter that uses a small shaped charge device rather than a mechanical cutter. Manufacturer is Rheinmetall Co.

Figure 5.9. Royal Navy "Minnow" unmanned Mine-Disposal Submersible. Has an unlimited dive performance capability.

thrusters and two additional lateral vertical thrusters for maneuvering when near a mine.

The underwater Remote-Controlled Vehicle (RCV) is not easy to design. It must have very low magnetic and acoustic signatures and practically no pressure signature or the mine will explode prematurely.

Tow-Taxi. (See Figure 5.10.) A towed underwater vehicle that can be fitted with side-scan sonar and other sensors and has mine-countermeasures capability. The manufacturer is British Aerospace Ltd. Normal operating depth is 100 meters, but a maximum of 250 meters can be sustained. Tow-Taxi was developed with MOD/UK funds and was delivered to Admiralty Research Establishment (ARE) at Portland, England for trials and operations.

Figure 5.10. British Aerospace has designed "Tow-Taxi," a towed underwater vehicle that can be fitted with side-scan sonar and other sensors.

CAT. Sperry Gyro Co. remotely controlled mine-disposal vehicle. (See Figure 5.11.) This is a catamaran which tows a weapon carrier. Attached to this carrier is an explosive charge, which is released by remote control from the mother ship, when over the mine. It is similar to the PAP-104 in concept. However, the PAP has proven more popular, so Sperry has had little success with the CAT. The CAT is powered by low-magnetic-signature diesel engines, each equipped with submersible mine-disposal weapons.

Trail Blazer—ROV system produced by Fairey Hydraulics Ltd. (See Figure 5.12.) The Trail Blazer has been jointly developed by I.S.E. Ltd. of Canada and Fairey Hydraulics of U.K. A very low-cost ROV that can be transported by land, sea or air to a craft of opportunity such as a fishing boat or harbor craft or used in MCMVs. The carrying container or van itself becomes a fully equipped operations room.

Has self-contained navigation, sonars, videos, side scans. Adapted for searching, identifying, marking, mine and torpedo recovery, mine disposal, security duties, and surveying. Can dive to depth of 500 meters, with a range of 1000 meters. Computer integrated with autonomous search vehicles, such as the Dolphin. Has good maneuvering capability in strong currents, in excess of 4 knots at depth. Has indefinite endurance, with only mine disposal charges needing reloading. Low maintenance, high reliability and availability are claimed. Naval trials were conducted in the Atlantic, Pacific and Indian Oceans, and in fresh water.

TUMS—A Royal Navy Towed Unmanned Submersible.

RCIV—Remote-Controlled Inspection Vehicle. Royal Navy intends to use for mine inspection.

ICMDS—Internally Controlled Mine-Destruction System.

Figure 5.11. British Royal Navy "CAT" Remotely Controlled Mine Disposal Vehicle.

Figure 5.12. The "Trail Blazer" minehunting vehicle. An inexpensive system that can be operated from small fishing vessels. Builder is Fairey Hydraulics Ltd. of Britain.

MICROV—Remote-controlled ROV built by Offshore Systems Engineering Ltd. (OSEL) and GEC Avionics of Britain.

SCARAB—A British mine-countermeasure ROV. It was developed by the Admiralty Research Establishment (ARE) for a cost of only $370,000. This is a small ROV to clear floating mines. It works by radio control and is guided to the target by binocular-enhanced visual means. When the SCARAB is within 50 meters of the mine, an onboard camera takes over and guides the craft in closer. Upon radio signal from the controller, the craft extends a grab which locks around the mine casing. When securely attached (the grab is designed so as not to detonate the mine by touching the horns), SCARAB releases the grab by firing a cable. The craft then moves away, and the mine is towed to a safe location and detonated by an explosive charge fitted to the grab. SCARAB then returns to its support ship for recovery and rearming.

MiniRover. A low-cost ROV built by Deep Sea Systems International. Purchased for use by the U.S. Navy EOD (Explosive Ordnance Disposal) Technology Center.

Sea Serpent—A minesweep system. Manufactured by Vosper Thornycroft U.K. Ltd. Has magnetic and can include acoustic sweep capability. Can be towed behind small craft of opportunity. (See Figures 5.13 and 5.14.) Utilizes microprocessor-controllable permanent magnets.

ARMS—Advanced Remote Minehunting System. A remotely operated vehicle to detect, classify, identify and dispose of all known types of mines to depths of 300 meters. Has 1000-meter umbilical cable to mother ship. Speed up to 6 knots. A new system being planned by Ministry of Defense (MOD) for the Royal Navy. A lightweight, portable system built to operate from a ship of opportunity. To detect, classify, identify and dispose of most known types of mines. Plessey Ltd. provides the

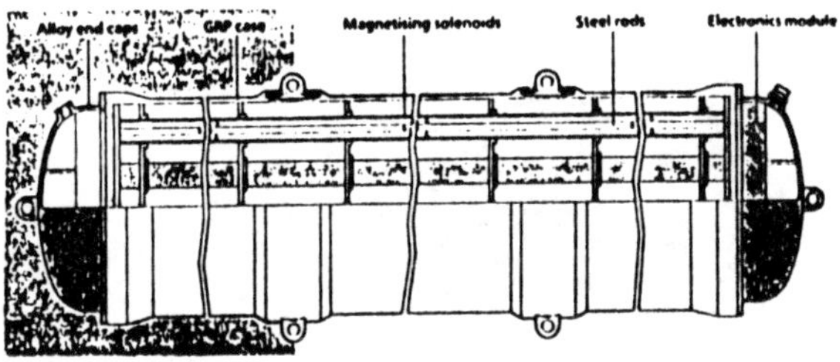

Figure 5.13. British "Sea Serpent" Mine-Countermeasures Device. Can be used for magnetic or acoustic sweeping.

Figure 5.14. "Sea Serpent" being towed behind small boat. The builder is Vosper Thornycroft of Britain.

acoustic mirror sonar and Ametek Co. of the U.S.A. provides Tandem Propeller System (TPS) technology. The ROV can maneuver in all six degrees of freedom or hover in a fixed position. The vehicle can be deployed up to 1000 meters in advance of the mother ship, at depths down to 300 meters. The Plessey sonar detects ahead at ranges greater than 600 meters and classifies bottom mines on mud/sand sea bed at a safe standoff range. ARMS can act as a picket to navigate a safe channel for a ship and following vessels, or dispose of mines by deploying a mine-disposal charge or explosive cable cutter.

Osborn—TAG (Towed Acoustic Generator). (See Figures 5.15 and 5.16.) Vehicle towed well astern of a vessel. It generates sound in a pre-selected wide-band frequency

Figure 5.15. The Royal Navy "Osborn" Acoustic Sweep System about to be deployed over the side of a British mine-countermeasure vessel.

Figure 5.16. A Towed Acoustic Monitor is a part of British Aerospace's Osborn Acoustic Minesweeping System.

spectrum. It also controls the acoustic amplitude and spectrum shape at the sea bed. Power and control cables are integrated into the main towing cable, which is supported from the surface with floats as is the TAG itself. A Towed Acoustic Monitor (TAM) is towed forward of the generator at a controlled height above the sea bed, monitoring the TAG-generated signal, thereby permitting control of the power output within discrete bands across the entire spectrum of generated frequency. This enables the desired acoustic signature to be maintained regardless of variable sea water characteristics, such as temperature, density or salinity, and ensures that the swept path width remains constant. Considerable power is used to ensure that the equipment can sweep a maximum swath width against varying mine sensitivities. A range of magnetic signatures can be simulated to represent the characteristics of a variety of surface vessels. Control of pulse timing and shaping is exercised from aboard the countermeasure vessel and allowance made for the type of mine to be swept and for the depth of water in which it is laid. The manufacturer of the "Osborn" system is BAe's Bracknell Division Ltd.

Oropesa Sweep—Conventional sweeping equipment for the clearance of moored mines. Manufacturer is BAe Dynamics. This system is comprised of a towed float, sinker and cable array that can be adjusted in depth so the cutter cable contacts the mooring cable of a mine.

Indonesia

The *Pulau Rupat*, a Tripartite-class mine-countermeasures ship built in The Netherlands, was recently added to the Indonesian Navy. This probably has a ROV such as the PAP-104 onboard.

Iran

The U.S. Navy sold the Shah of Iran six systems of helicopter-towed mine-countermeasure systems. These include the Sikorsky Co. "Sea Stallion" helicopters and the EDO Co. Mk-105 towed sled systems. The present regime in Iran should still have these in their inventory.

Israel

The Israeli Navy has obtained some ROV craft for various purposes, including mine-neutralization operations.

Italy

Pluto—Italian Navy MCM ROV system. Eight will go into newly built "Lerici" class MCM ships. The other will be installed on the rescue-diving vessel *Anteo*. Plutos are installed on all operational "Lerici" class MCM ships, some of which have operated in the Persian Gulf. (See Figures 5.17 through 5.21.)

The manufacturer is Gaymarine, Trezzano sul Vaviglio Co. of Italy. Westinghouse, Oceanic Division of the U.S.A. has been working with Gaymarine on the Pluto. The vehicle can carry a mine-disposal charge to blow up mines.

Gaymarine has developed the Pluto Plus ROV system. It is designed for operation in strong currents and gives extended submerged endurance. The vehicle weighs 200 kg, measures 2 meters long, is depth rated to 400 meters and carries

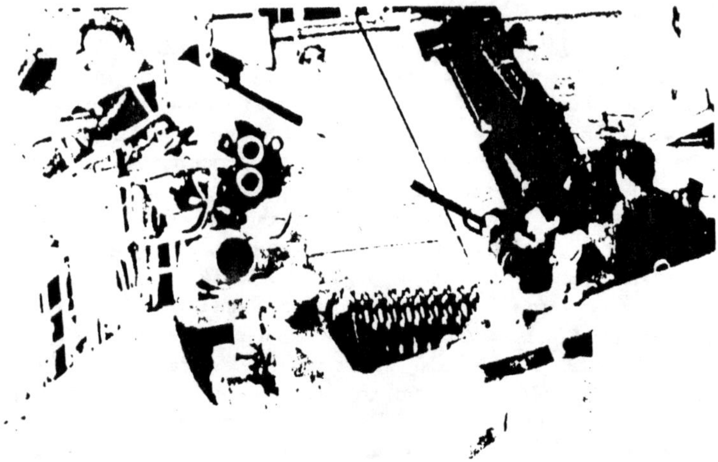

Figure 5.17. The Italian Navy "Pluto" with mine-disposal charge, being launched on a "Lerici" class MCM ship.

Figure 5.18. The "Pluto" can cut cables or place explosives to neutralize mines. It carries sonar for target classification.

Figure 5.19. Gaymarine Co. of Italy "Pluto Plus" ROV System on sea trials for the Italian Navy.

Figure 5.20. Forward view of "Pluto Plus." Designed for strong currents and extended endurance.

Figure 5.21. Italian Gaymarine Co. recently introduced a small-size "Pluto" Mine-Disposal Vehicle.

Figure 5.22. Italian Navy "Lerici" class GRP Minehunters use the "MIN" Mine-Disposal Vehicle. The builder is Riva Calzoni Co.

a color TV camera and sonar within a wet, transparent dome. It is battery powered, with submerged operation rated at 2 hours with high speeds at about 7 knots. The battery capacity is double that of the standard Pluto. The umbilical cable is 3 mm diameter fiber optic cable supplied in a length of 2 km. The television camera and triple sonar suite (for search, navigation and identification sonar) are mounted in a single package on a ±80-degree pan and ±100-degree tilt unit. Automatic heading and depth controls are included.

The Pluto has also been purchased by other navies.

MIN—Mine Disposal Vehicle. (See Figure 5.22.) Built for the Italian Navy and used on "Lerici" class MCM vessels. A joint development of two Italian companies; ELSAG S.p.A (Elettronica San Giorgio) and Riva Calzoni S.p.A. The two firms are known as SMIN concern.

The system known as MIN-77 is similar to the French PAP-104 system but has some important differences:

- Propulsion motors are hydraulic, fed by pressurized fluid.
- Very low acoustic and magnetic signatures.

- Equipped with TV camera and/or a high-definition sonar.
- Can be controlled either by a main console or by a small portable console.
- Specifically designed for disposing of either bottom or moored mines.

Japan

The Japanese Maritime Self-Defense Forces have a number of minehunting vessels on which ROV units are deployed. Their Navy has also purchased some minesweeping helicopters from Sikorsky Co. in the U.S.A. that can deploy remote sweeping devices.

Korea (South)

The South Korean Navy uses ROVs for military purposes. As they are building minehunting vessels at the Swallow yard, it is conceivable that the ROVs could be deployed from the new minehunters.

Malaysia

The Navy of Malaysia is known to have remote-operated mine-clearance vehicles.

Netherlands

The Royal Netherlands Navy uses the PAP-104 French mine-neutralization system on their "Tripartite" class MCM vessels.

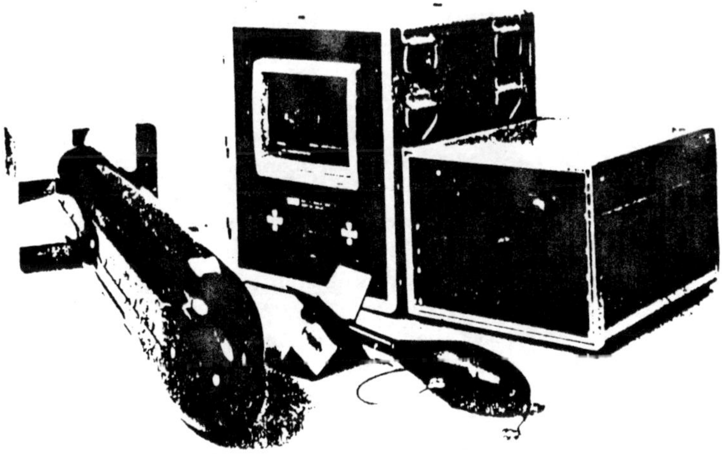

Figure 5.23. Model 972 towfish and towfish depressor, display monitor and side-scan sonar processor, remote winch controller and video data recorder/playback. Simrad Group of Norway builds this system at Mesotech Ltd. in British Columbia, Canada.

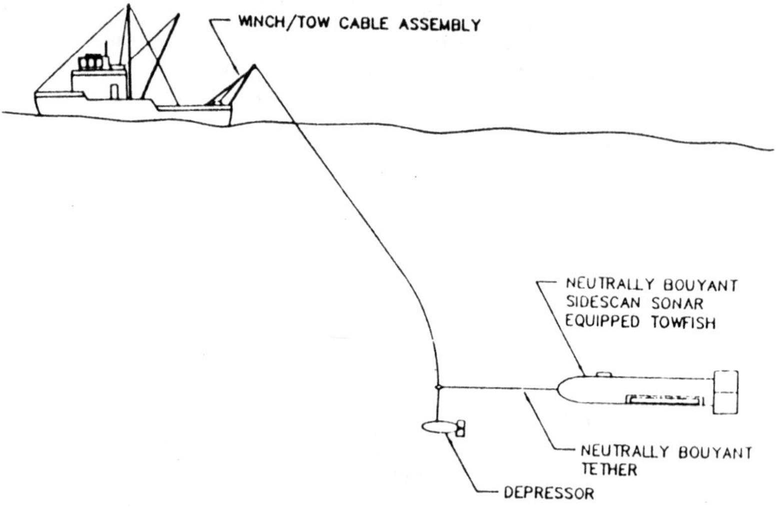

Figure 5.24. The Mesotech Systems towfish decoupling method.

Nigeria

The Navy of Nigeria utilizes foreign-purchased ROV vehicles.

Norway

Model 972. The Simrad Group of Norway purchased the Mesotech Co. of British Columbia, Canada. With it came the Model 972 Side-Scan Sonar and bottom profiling system with color display. (See Figures 5.23 and 5.24.)

This is primarily a search vehicle and does not have the ability to carry an explosive charge such as would be needed to clear a mine. However, it is mentioned here under towed, remote vehicles because the technology is appropriate. A unique towfish decoupling method is used, with a depressor. The fish is thus decoupled from surface sea state action by use of a weighted depressor and a neutrally buoyant tether attached to the armored cable.

The towfish is rated at 1000-meter depth, measures .15 meter in diameter, about $1\frac{1}{2}$ meters in length, and weighs 15 kg in air. The Model 972 incorporates two wire telemetry and power supply systems to the fish, thereby easing tow cable requirements. The side scan operates at 120 kHz and 330 kHz. The sub-bottom profiler has a 10-kHz operating frequency, with primary frequencies of 200 and 210 kHz. A low-frequency pulse is generated using a parametric technique, by transmitting two high-frequency pulses and combining them in the water. This produces a narrow, well-defined, low-frequency beam.

Sweden

Sea-Owl. Swedish ROV. Used in 1982 to inspect and videotape the Soviet Whiskey submarine bottomed out for several days in Swedish waters offshore from Karkskrona Swedish Navy base.

Figure 5.25. "Sea Eagle," a new unmanned mine-disposal system from Sutec Co., Sweden. The long arm can be extended with a TV camera attached to obtain close-up pictures of a mine. This system was used by the British Royal Navy to locate a Soviet-made mine in the Middle East.

Figure 5.26. Swedish Royal Navy "SAM" Self-Propelled Remotely Operated Acoustic Magnetic Minesweep.

Sea-Eagle. A new Swedish unmanned mine-disposal system from Sutec Co. Has a long arm that can be extended to obtain close-up pictures of a mine. This system was used by the British Royal Navy to locate a Soviet-made mine in the Middle East in the 1980s. (See Figure 5.25.)

SAM – Swedish Self-Propelled Remotely Operated Acoustic Magnetic Minesweep. Manufacturer is Karlskronavarvet of Karlskrona, Sweden, a member of the Celsius Group. (See Figure 5.26.)

U.S.A.

MNS (Mine Neutralization System). The JAN nomenclature of the system is ALQ-48. Vehicle power is provided through a neutrally buoyant umbilical cable, allowing the MNS to take on extended-duration subsea tasks, including:

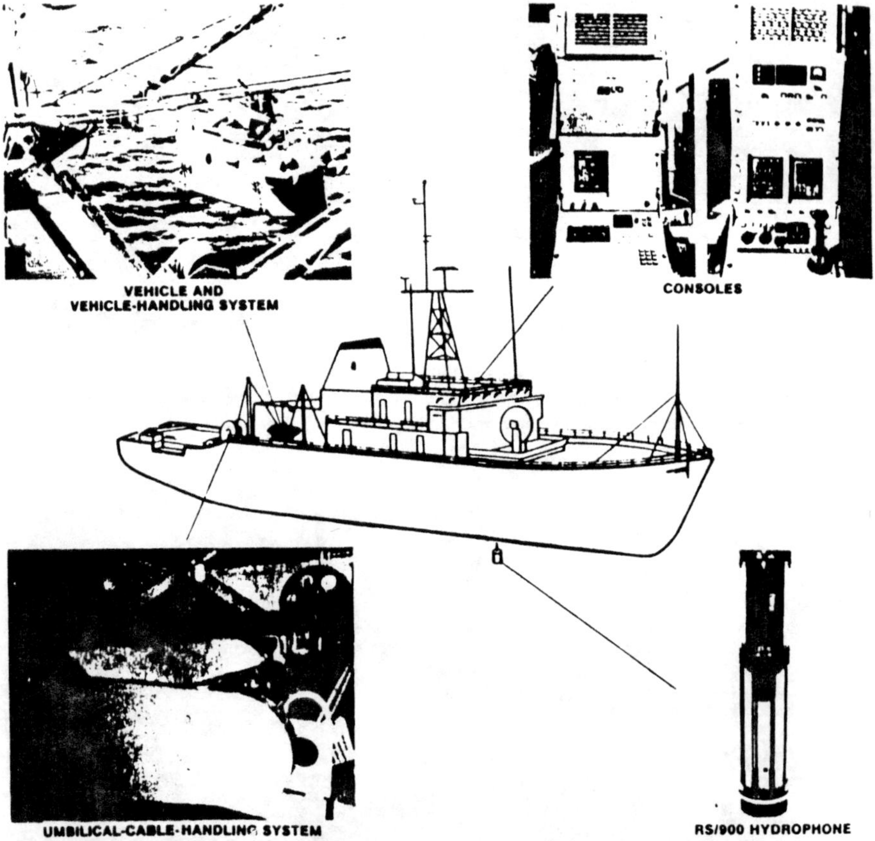

Figure 5.27. The U.S. Navy MNS (Mine-Neutralization System) components as installed on a typical minehunter. Honeywell is the manufacturer. System is called the SLQ-48. Effective shipboard integration will give high performance and versatility. Note the diagram below.

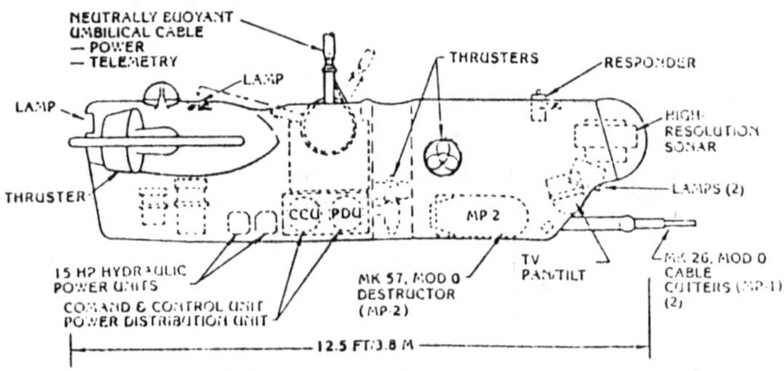

- Channel conditioning
- Path finding
- Subsea inspection
- Object recovery
- Mine disposal

See Figure 5.27 for a view of system components as they are placed on a typical minehunting vessel. Figure 5.28 and 5.29 show the rather complex vehicle-handling system and the operating consoles.

The sonar screen displays sensor and vehicle-status information in digital form. The monitor and control consoles are the central point for all system-control functions, including:

- Performance monitoring
- Fault location
- Sonar
- Television
- Secondary sonar and TV presentations
- Vehicle control
- Cutter and dropped-charge actuation

Figure 5.28. SLQ-48 Mine-Neutralization System Handling System. It carries lights, TV cameras and sonar. Controlled from the ship. Can drop an explosive charge near a mine. Note the complex handling system.

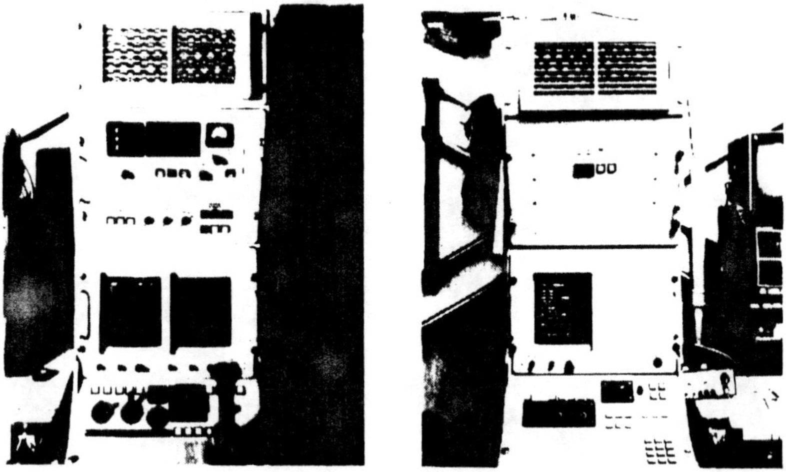

Figure 5.29. Control console (left) and monitor console (right) of the SLQ-48 MNS System.

Initial target detection and vehicle-guidance information is provided by ship sonar. Vehicle tracking is also provided by a Honeywell RS/900 Acoustic Position Indicator. Vehicle sonar is used during midcourse-search and final vehicle-guidance phases. High resolution provides the capability to conduct missions in turbid waters under sonar guidance alone. Low-light TV is used with sonar during the on-scene, precise maneuvering phase. Underwater launch and recovery supports operation in high-sea states. A computer-generated vehicle-navigation track ensures the fastest possible mission time.

The MNS was originally designed by Naval Ocean Systems Center (NOSC) in San Diego, California. The Advanced Development Model (ADM) was refined under a competitive contract by Honeywell Co. and Hydro Products Co. into two Engineering Development Models (EDM). The two companies then teamed (Honeywell purchased Hydro Products) to win the production contract. Honeywell Undersea Systems Division, Hydro Products Operations produces the MNS vehicle. When completed in 1989 (?), Honeywell will have supplied 14 systems and 15 spare MNS vehicles for oceangoing MCM ships and coastal MHC ships. The first 9 vehicles were produced for the U.S. Navy at a cost of $30 million.

The MNS vehicle emits low acoustic and magnetic signatures, enabling it to do its job undetected by mines. Twin hydraulic longitudinal thrusters, along with vertical and horizontal thrusters, permit delicate maneuvers close up, even in moderate currents. A camera, housed in a bubble on the top, aft area of the vehicle, is used during launch and recovery to ensure that the vehicle is properly attached to the handling system. The camera also provides the operators with a clear view of the umbilical cable, allowing them to correct any tangles and twists. Needless to say, this is an elaborate and complicated handling system.

The MNS carries two cable cutters for moored mines and a destruction charge for bottom mines. Once these are used, the crew must recover, rearm and relaunch the vehicle for its next mission. A fiberglass frame and skin-housing sections of syntactic foam give the vehicle positive buoyancy. In an emergency, such as cable tangling, the operator can release a lead ballast from the vehicle and its umbilical is automatically severed. The vehicle surfaces and an emergency pinger and flasher are activated so it can be located and recovered.

A few of the MNS specifications are as follows:

Dry weight	1130 kg (In water: 9 kg positive buoyancy)
Speed	6 knots
Length	3.8 meters
Height	1 meter
Propulsion	15-hp twin hydraulic system
Cable length	1000 meters
Tensioning	10-300 kg
Reel speed	0-200 meters per minute
Weight	10,000 kg (total system)
Power reqmt	60 kW peak
Maintainability	4 hours corrective maintenance time
Reliability	120 hours MTBF (Mean Time Between Failures)
Test	Built-in Test

The MNS can locate, classify and neutralize bottom and moored mines. It utilizes a microcomputer command and control system.

Super SeaRover. Manufactured by Benthos Inc. of North Falmouth, Massachusetts. An improvement over the two previous ROVs produced by the company; SeaRover, a militarized version of Low-Cost ROVs (LCROV), the MiniRover Mk I and Mk II. A recent U.S. Naval Sea Systems Command $2.5 million contract was completed for eight of the Super SeaRovers. They will be used aboard U.S. Navy minesweepers. (See Figure 5.30.)

This ROV is powered by six variable-speed thrusters. It is equipped with a low-light color TV, a scanning sonar and an acoustic tracking system. With the Benthos

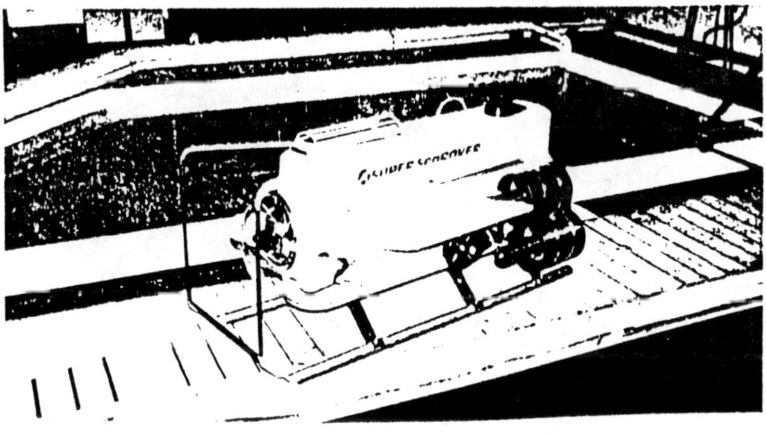

Figure 5.30. Super SeaRover - a militarized version of Low-Cost ROVs. Produced by Benthos Inc. for U.S. Navy minesweepers.

Super SeaRovers, Navy divers will no longer have to go out in rubber rafts to examine suspicious items that could be mines. Originally labeled Standard Navy Vehicle (SNV), the new Super ROV features improvements over the SeaRover. The ROV has a very low magnetic signature. Good control of vehicle movement is achieved by stacking six pairs of horizontal thrusters, which give it a speed of 4 knots underwater, or 2 knots against an opposing 2-knot current. The brushless motors deliver a thrust of 45 kg. The vehicle has a weight of 70 kg. Standard equipment onboard is:

- Mesotech high-resolution scanning sonar
- Ferranti O.R.E. Trackpoint integrated navigation system
- Hitachi high-resolution CCD Video Camera mounted on a pan and tilt
- 450-meter umbilical cable—Super SeaRover rated depth is 350 meters
- Video graphics overlay package
- Beta-format video recorder

Klein Mk24 Side-Scan Sonar for MCM applications. U.S. Navy EOD (Explosive Ordnance Disposal) facility procured 16 of these sonars under a $1.5 million contract. (See Figure 5.31.)

The Mk24 is comprised of:

- 100-kHz 3/4-degree beamwidth side-scan sonar towfish
- K-Wing I hydrodynamic depressor
- Tow cable
- Two-channel transceiver
- Graphic recorder

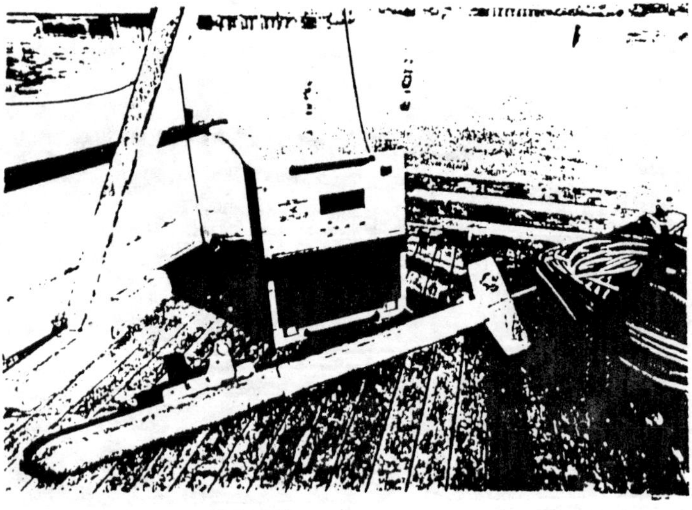

Figure 5.31. Mk-24 Klein Co. Side-Scan Sonar Towfish with other equipment for MCM operations.

- Alphanumeric annotator providing date, time, event, and navigational data on the side-scan sonar
- Options include a 3.5-kHz sub bottom profiler and a 50-kHz long-range side-scan sonar

This is primarily a sonar and there is no provision to carry an explosive charge (as in the larger ROV systems) to clear a mine. The Klein Co. is located in Salem, New Hampshire. Klein equipment is in use by clients in over 40 countries as well as in over a dozen U.S. Navy agencies.

Phantom—Spectrum series of ROVs for MCM, AMCM. Produced by Deep Ocean Engineering Co. of San Leandro, California. (See Figure 5.32.)

Scorpio—ROV with four thrusters to control depth and heading. Has a five-movement manipulator, and 250-watt tungsten-quartz and thallium iodide television lamp.

LENS—A low-cost underwater vehicle that can be launched from helicopters to destroy enemy naval mines. It is wire-guided and battery-powered, and is called the Low-Cost Expendable Neutralization System. Lockheed Missiles and Space Co. of Sunnyvale, California was the builder for the U.S. Naval Air Systems Command. A contract for $5.2 million included the delivery of 100 test models and two remote-control consoles. This is a follow-in to the LENS system.

ADROV—Advanced Development ROV. A prototype that was developed by Hydro Products Co. for the U.S. Navy's Explosive Ordnance Disposal Technology Center.

BURMMS—Buried Mine Minehunting System. Under development for U.S. Navy by Westinghouse Electric Co.

VIPER Mk 1—A high-speed robot undersea vehicle designed for waterside security, mine countermeasures and channel-conditioning missions. (See Figure 5.33.) The manufacturer is Allen Osborne Associates of Westlake Village, California.

The vehicle carries a high-resolution sonar system which has a range of several hundred meters, a low-light CCD video camera, and can carry a variety of payloads

Figure 5.32. Phantom ROV for MCM, AMCM and other Naval requirements.

Figure 5.33. VIPER Mk-1 High-Speed Robot Undersea Vehicle. Has a high-resolution sonar and low-light CCD video camera. Manufacturer is Allen Osborne Associates.

including ordnance-neutralization charges and intruder-response systems. It is propelled by four high-power brushless, DC reversible motors, and has a forward speed of 8 knots and reverse speed of 5 knots. Maximum operating depth is 200 meters. The vehicle length is 1.88 meters, width is 0.5 meters, and height is 0.3 meters. Weight is 46 kg. The main body is made of extruded PVC tube and tailfins and nacelles of kevlar/epoxy laminates. The tether cable is fiber optic, fiberglass reinforced with break strength of 150 kg and is drawn from a canister in the rear of the vehicle. Operational radius is 2 km. The operator uses a joystick and mouse to control the Viper, with a heads-up display. Control is augmented by the vehicle's auto-heading, depth and altitude functions.

Mk-105 – U.S. Navy Mine-Countermeasures System. (See Figures 5.34 through 5.37.) Use catamaran hulls. Utilize acoustic and magnetic sweeps to clear both moored and bottom mines. Extensive experience was gained by the U.S. Navy in the waters of North Vietnam in 1973 when the Navy was given the responsibility for clearance of mines that it had laid as part of the blockade measure enforced on North Vietnamese ports. Large Sikorsky Sea Stallion helicopters were used to tow the equipment.

An electric generator powered by an engine on the Mk-105 sled provides electrical current to sweep mines. The sweep operator, at a control console in the helicopter, can start and stop the turbine on the sled and raise and lower the

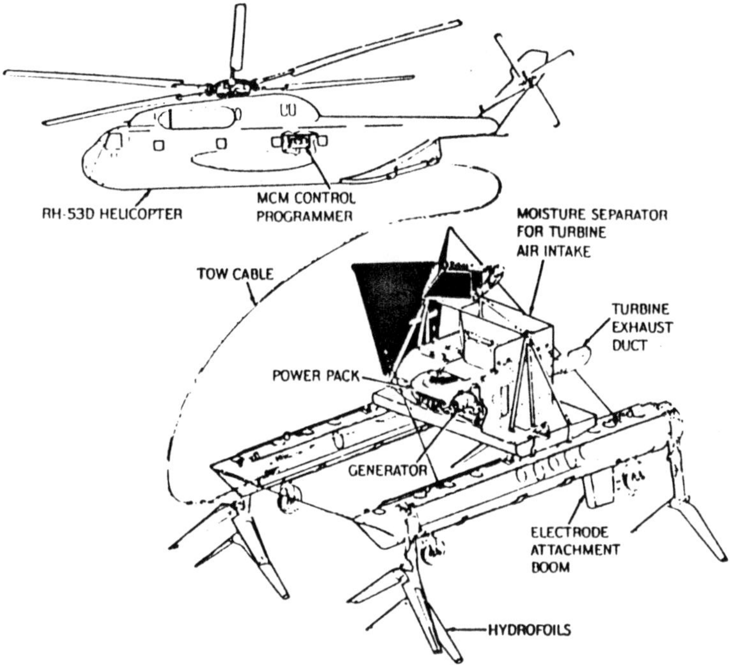

Figure 5.34. RH-53D U.S. Navy helicopter with elements of the Edo Co. Mk-105 mine-countermeasures system.

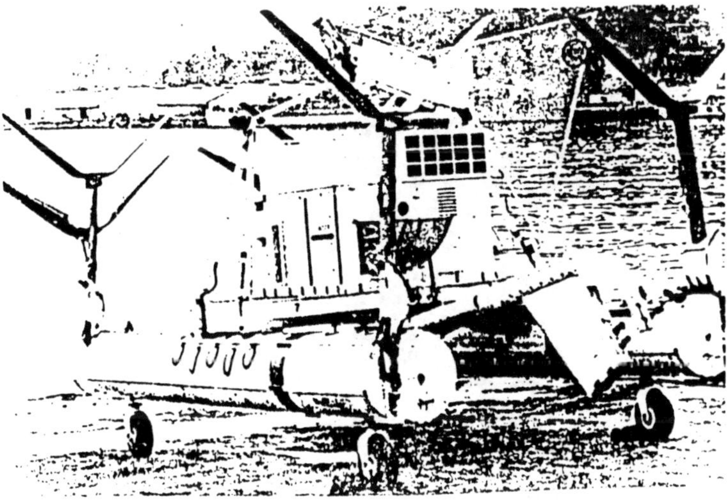

Figure 5.35. Close-up of the Mk-105 Magnetic Hydrofoil Minesweeping Sled.

Figure 5.36. U.S. Navy Mk-105 Mine-Countermeasures System being lifted off the deck of a Navy ship.

Figure 5.37. The Mk-105 MCM system shown being towed. The sled can operate three to four times faster than conventional minesweeping ships with little hazard to the helicopter crew.

hydrofoils. He can also control the output signals that create the magnetic field. It can be constant, or it can be pulsed so that its rate of build-up and decay simulate the mine's target ship.

The foil-borne sleds have a better chance of survival when the mine detonates than would a surface craft of conventional construction. When a mine blows, the craft moves a bit.

Oropesa—Sweep Equipment. Conventional sweeping equipment for the clearance of moored mines. Similar to British system manufactured by BAe Dynamics. Has a towed float, sinker and cable array that is adjusted in depth so the cutter cable contacts the mooring cable of a mine. Can be towed by helicopters such as U.S. Navy Sea Stallions.

MOP—Magnetic Orange Pipe. (See Figures 5.38 and 5.39.) These devices are towed behind helicopters such as U.S. Navy or Marine Corps Sea Stallions to magnetically clear mines.

ALQ-166—U.S. Navy Helicopter-towed Mine-Countermeasures Vehicle. To lift the sled, the helicopter dangles a ball on a line and maneuvers it into a V at the top of a tripod-like structure. (See Figure 5.40.) An improvement over the older Edo Co. Mk-105 sled. Has monohull-type construction, can carry a heavier load—more fuel for the 1200-hp Avco Lycoming engine, which is about three times as powerful as the one carried by the Mk-105. The larger engine drives a higher-output generator and enables the magnetic sweeping equipment to sweep a broader path and detonate mines in deeper water. The heavier, larger ALQ-166 was designed to be towed by the new, more powerful helicopter called the MH-53E Super Stallion.

Figure 5.38. A U.S. Marine Helicopter is towing a MOP (Magnetic Orange Pipe) during Operation "End Sweep" 1973 off North Vietnam.

Figure 5.39. Crewmen on *USS Ogden* waiting for RH-53 helicopter to take MOP under tow. The *Ogden* is an LPD which has a boat well aft on the ship, which facilitates the deployment of the MOP.

Figure 5.40. U.S. Navy ALQ-166 Helicopter-Towed Mine-Countermeasures Vehicle.

Thailand

The Royal Thailand Navy utilizes some ROV craft for mine countermeasures. These craft have been procured from foreign countries.

U.S.S.R.

SPA—Self-Propelled Underwater Apparatus. Similar to remotely controlled submersible vehicles in U.S. Navy and other Western navies, such as PAP-104 or MNS.

Ilyusha—Class of minesweeping boats that are radio-controlled minesweeping drones. Controlled by a mother ship such as Vanya minesweepers. A system somewhat similar to the West German Troika.

Possibly new 40-meter remote-controlled unmanned minesweepers under construction.

Soviet helicopters probably have a magnetic/acoustic mineclearing system similar to the U.S. Navy Mk-105 and ALQ-166 systems.

Soviets acknowledge that they have produced at least one 1500-meter operating depth ROV.

Additional General Information

European-based Scicon Co. has conceptually developed "SPUR"—Scicon's Patrolling Underwater Robot. The full-scale model would be between 10 and 11 meters long with a 1.8-meter cross-section. Normal propulsion would be by an oxygen/hydrogen fuel cell, and higher attack speeds of up to 50 knots may be obtained with either a closed-cycle engine or batteries.

At cruising speed, the vehicle would have an action radius of 1850 km, an endurance of two months and, for certain applications, be capable of diving to 6000 meters. SPUR would be under shore control and artificial intelligence would be used to assist functions which SPUR would need to perform autonomously, especially in tactical decision-making, route planning, target classification, attack maneuvers and communications routines.

The Scicon Co. seeks such roles for SPUR as mine countermeasures, using manipulators, vessel destruction via torpedoes, wrap-around wire system deployment directed against a ship's propellers and, as a last resort, an intelligent mobile mine. The company envisions "wolf packs" of SPURS deployed in a patrol line to create barriers up to 370 km long.

Gaseby Dynamics Ltd. of Britain is currently developing an expendable intelligent decoy. It is mounted in magazines on a surface vessel. As many of six of these vehicles can be deployed into the water if the presence of acoustic torpedoes is suspected. Note: These torpedoes could originate from an underwater waiting mine such as the U.S. Navy Mk-60 Captor. Against a passive seeker the decoy would automatically begin a deduction or distraction program, while against an active seeker it would store sonic transmission, then retransmit it with added doppler and time difference to seduce it also. Such decoys could also be launched from aircraft or submarines.

Submarines could easily launch remote scout vehicles (ROVs) to check ahead of the submarine for mines. Avoidance maneuvers could then be taken by the submarine or action taken to destroy the mines.

CHAPTER 6

Mine Warfare Navigation/Plotting/Display Equipment

GENERAL

A requirement for those involved in mine warfare is to have good navigation accuracy when making route surveys, conducting channel conditioning along strategic routes (Q-routes) and planting or sweeping mines. This is true for aircraft, surface ships and submarines.

It is vital that the position of all mines planted or found should be accurately established and recorded. Modern minehunters are therefore now fitted with a complete action information system that includes displays of the above-water and below-water tactical situation. Similarly, they must have a highly accurate navigation system and particularly a position-fixing system accurate to a few meters.

A number of such systems have been developed, all working on the hyperbolic principle. The French have two systems, Toran and Syledis; the U.S.A., Royalist and Argo; and the British, Racal Hi-Fix/6. Typical operation is from a number of radio transmitters onshore creating position lines of known phase difference on the hyperbolic principle. The ship receives these position lines and locates herself on at least two of them. Where the lines intersect is the ship's position. The position of the mine is determined by a bearing and range obtained by sonar from the ship. The shore stations can be portable and can be moved as the mined areas are altered.

Manufacturers are offering a complete minehunter package. For example, the French have the Thomson-CSF IBIS system, which consists of:

- A minehunting sonar (TSM 2022)
- A tactical display console
- A doppler sonar log
- Navigation systems, autopilot, windspeed sensors, etc.

MAINS

The British have a complete action information system, called MAINS. This stands for Minehunting Action Information and Navigations System. The manufacturer is Racal Ltd. It is purely a command and control and navigation system and consists of:

- High accuracy navigation by radio aids, such as Hi-Fix/6
- Ship control guidance for all phases of minehunting
- Integration with any minehunting sonar
- A combined surface and sub-surface situation display
- An automatic plotter and data printer
- An automatic chart table
- Data recording and retrieval

The MAINS comprises an integrated family of navigation and radar sensor, minehunting sonar, data processing, display and recording equipment that may be combined in a number of ways to provide a coordinated tactical plotting system for minehunting. The system is computer controlled and provides a PPI (Plan Position Indicator) type presentation of the combined surface and sub-surface tactical situations to the minehunting control officer.

Ship's position information can be obtained through any one of a number of high-precision navigational aids such as Hi-Fix, SatNav, Loran-C, etc., and accurate dead-reckoning is derived from doppler sonar and gyro-compass inputs or an EM log. A secondary positioning mode may also be derived from the vessel's navigation radar to relate position to known radar contacts such as buoys. The system provides ship control guidance for general navigation, accurate track keeping to ensure full sonar coverage, and maneuvering around sonar contacts to hold the vessel at a designated position during mine disposal. An autopilot may be directly controlled by the MAINS computer. Sonar integration with the system offers three principal advantages:

- Optimal search coverage
- Designation of targets with immediate calculation of contact position
- Presentation on action information displays

MAINS also provides sonar control for search of selected sea bed areas and can hold a contact during classification and disposal. It can be integrated with any modern sonar system.

In operation, the ship's surface position relative to the sonar search plan and sub-surface targets are presented together on a single PPI. A large-scale, alpha-numeric data display provides information retrieval, and facilities are also provided for automatic hard-copy plot printing. Magnetic tape cartridge recording of sonar information is provided for either subsequent onshore analysis or as background information for future sonar search of an area. Although MAINS is a most comprehensive minehunting control facility, it may be operated by one man.

CAAIS

Ferranti Ltd. Computer Assisted Action System. Another example of the application of systems integration for minehunting and neutralization. It is a command and control action information center for larger classes of vessels, but a special version was developed for minehunters. For the safe hunting of magnetic mines, a very low magnetic signature is required. If electronic equipment can be kept small and as non-magnetic as possible, this is an advantage to the minehunter. This Ferranti Ltd. system is used on the Royal Navy Vosper Thornycroft Hunt class mine ship.

MCCS

This is the smaller Ferranti Ltd. Mine Countermeasures Control System mentioned above. It is a two-man-operated system but is equipped with a more powerful computer, the F1600E. Radio aids are used more than radar in this system. The tactical plot utilizes computer-generated symbology derived from the sensor and navaid inputs and uses a "prompt" system to lead the operator through his list of available operational procedures and options.

RACAL 880

Another British system is the Racal 880. It is a form of action information system for use by minehunters. It integrates the sonar with the navigational system and accepts radar information.

SERPENT

Serpent, another British system by Vosper Thornycroft, is intended for use against magnetic mines. It consists of a number of towed bodies containing a permanent magnet, the signatures of which can be modified from the towing vessel. The signature of the towed bodies can be set to zero during recovery or mooring. For sweeping a channel, two bodies are towed spread out on each quarter of the towing ship and programmed to produce a particular type of magnetic field between them, which is used to explode the mines below. Alternately, six bodies can be towed, one behind the other, each programmed to produce a magnetic field of a different ship, such as frigate, cruiser, aircraft carrier, etc. The system can be set to sweep mines that are adjusted to explode on a certain number of counts or vessels passing. This system needs good navigation and positioning accuracy.

SSN-2 PINS

The AN/SSN-2 Precise Integrated Navigation System (PINS) used aboard the U.S. Navy's MCM ships will permit the ship to perform her duties anywhere in the world and without control of the coastline where it is operating.

The PINS uses a powerful minicomputer, and it performs many command and control functions in addition to integrating the navigation data into an optimal solution. The SSN-2 is manufactured by Magnavox Marine and Survey Systems Division.

The navigation suite in the ship should consist of:

- Conventional speed log and gyrocompass for dead reckoning
- Land-based radio navigation for precise positioning during mission
- Global or positioning system

Short-range systems operating in the 5- to 9-GHz frequency range, such as Motorola Mini-Ranger and Del Norte Technology Trisponder provide good accuracies over relatively short ranges; i.e., line of sight or up to 80 km.

The new GPS (Global Positioning System) may negate the need for land-based systems. This requires a sufficient number of satellites for continuous two-dimensional coverage.

Precision doppler sonar speed logs and gyrocompasses can be added to the system to improve the dead-reckoning accuracy of the system. When calibrated, accuracies of 0.5% to 1% of distance traveled are achievable. When used for precise navigation, these systems include a satellite receiver to reset the time varying error growth of the doppler sonar/gyrocompass. One can expect the error growth of such a system to remain between 50 to 100 meters as long as the doppler sonar is in the bottom track mode; i.e., receiving signals reflected from the ocean floor.

HYPER-FIX & QX-3A

Racal Ltd. of Britain has sold the U.S. Navy some Hyper-Fix and QX-3A precise navigation equipment for use in MCM and related operations. The Hyper-Fix includes shore stations, shipboard receiving and QX-3A track guidance systems, and MCM helicopter receivers. The Hyper-Fix 2-MHz phase comparison precise positioning system is claimed to provide remote-control facilities and unambiguous positioning accurate to within five meters with a range of 250 km. Racal's QX-3 MCM information system is intended for both dedicated minesweeping vessels and minehunters. Hyper-Fix and QX-3 systems are now standard equipment for MCM for the Royal Navy, U.S. Navy and other NATO navies. AIS is meant to indicate an Action Information System that fulfills all the requirements of minehunting and hydrographic survey.

QUILS II

The Royal Australian Navy has selected a Meridian Ocean Systems QUILS II Navigation and Data Management Package for their Craft-of-Opportunity Mine Surveillance System. QUILS provides a single integrated system which allows an operator to conduct high-accuracy side-scan sonar surveys of the sea floor, digitize map coordinates of important sea floor targets as they are observed in real time and analyze all target information (current and historical) via an interactive color graphics console.

Meridian Ocean Systems Co. is currently adapting the flexible QUILS II architecture to a variety of tasks including sea floor mapping, hydrographic surveys, search and recovery and ROV operations.

NAVIPLOT TSM-2060

The French Navy NAVIPLOT TSM-2060 tactical plotter is used on their MCM vessels for determination of the precise geographical position of target mines and the recording of other relevant data. The NAVIPLOT is based on the CIMSA 15MO5 Processor. Information is displayed on four color cathode ray tubes on which raw video can be overlaid in graphic symbology. The French company Thomson-CSF system related to this is the "IBIS."

It is vitally important always to fix the position of any mine detected as accurately as possible. One reason is that the mines may have to be left for a while in their position and not exploded or removed immediately. The weather could have deteriorated or the mine might have been located just before dark. Another reason is that the limits of the minefield must be determined so that shipping can be warned to keep clear of it. During World War II, and until quite recently, the position of a mine was marked by buoys and its geographical position plotted. It was not a very accurate method and valuable time was often lost on return to the position of the buoy in trying to locate the actual mine again. In addition, the laying of the buoys and the fixing of their positions was a time-consuming job.

Radio navigational aids have now changed this, and improved navigation and recording of mine locations is now possible. The hyperbolic system is most widely used. An MCM vessel, equipped with the required receiver, notes her own position and also the position of the mine in relation to her by sonar bearing and range. When the vessel returns to the area, she is supplied with the position of the mine, either in latitude and longitude or as a grid reference, and can go straight to the position and find the mine without difficulty. Accuracy is very important, since if the mine is to be destroyed, the vehicle which lays the destruction charge cannot waste time in searching for the mine, a job which the ship has already done.

There are various radio position-fixing systems. These include the Hi-Fix/6, Hyper-Fix, Microfix and Trisponder, all of which can provide an accuracy to within a few meters. The shore transmitting stations for creating the hyperbolas can be set up permanently in known areas of enemy minelaying, or they can be transportable and placed in the positions required onshore in semi-permanent sites. They can even operate unmanned. Onboard the minehunters, there is a receiver to receive the signals from the shore stations and a computer to work out the positions of the mines from sonar bearings and ranges. However, coming more into use these days are automatic systems which correlate the sonar and navigational data and control the whole system.

MWS-80

The German Navy has purchased the MWS-80 and DSQ-11H sonar from the Krupp Atlas Elektronik Co. of Bremen, West Germany for MCM operations. Some of these systems have also been sold to foreign navies such as Australia and Thailand. See Figure 6.1 for a view of the operational console inside the MWS-80 mine-countermeasures van.

This is essentially a system that correlates the sonar and navigational data and controls the whole system. The MWS-80 is an integrated, multi-role mine-countermeasure system that performs search, detection and classification of mines, as well as maintaining full control and coordination of all MCM operations. It maintains a precise record of all areas searched and mines located, together with their accurate locations and classifications. The system includes the minehunting sonar DSQ-11H and the navigation equipment. It can be integrated with other high-precision navigational systems, the primary system used by Atlas Elektronik being the Motorola Miniranger MRS or the Racal Ltd. Trisponder.

RADS

The Racal Ltd. firm of Britain makes a system similar to the MWS-80, or rather a combination of two systems; the Racal Integrated Navigation System and the Racal Action Data Display System, collectively known as System 880. See Figure 6.2 for a view of the Racal Action Display System (RADS).

It can use all of the standard radio navigation systems. The previously mentioned Action Information System (AIS) provides full interfacing of all sensors, data presentation, and processing and logging in all modes of operation, including hydrographic survey, route survey and mine location and clearance.

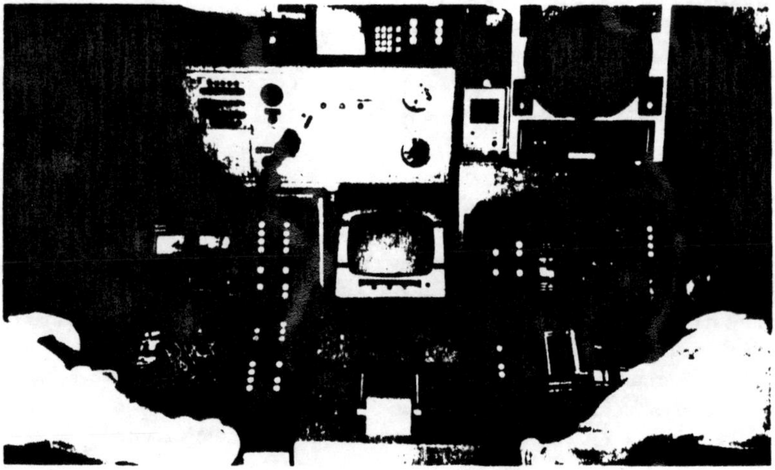

Figure 6.1. German MWS-80 Mine-Countermeasures System Operating Console installed in a portable shelter. Manufacturer is Atlas-Elektronik. This system is in service with West German, Australian and Thai Navies.

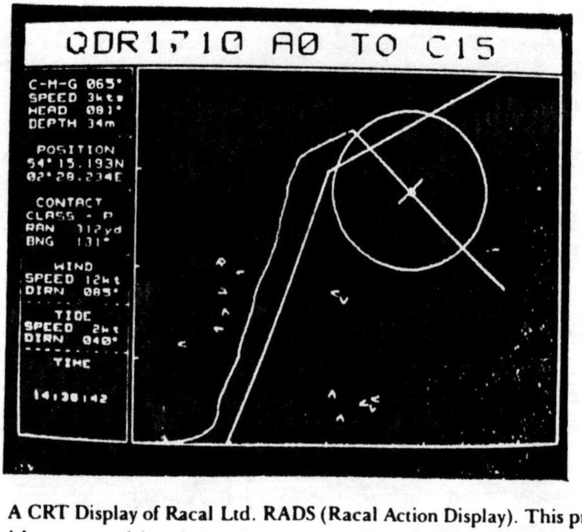

Figure 6.2. A CRT Display of Racal Ltd. RADS (Racal Action Display). This provides contact classifications with range and bearing.

In the hydrographic survey role, digitized soundings from an echo sounder are sampled and logged together with position. The sonar is used to provide an initial map of the sea bed, noting any changes in texture and significant features.

The off-line post processing of the data can be carried out ashore, using a system similar to that on the vessel. As well as the fair sheet and all accompanying tracings, the sea bed map made up from sonargraphs, and pseudo-three-dimensional digital elevation models, are produced. These charts, when considered together, provide an accurate and concise presentation of all the information needed for planning of Q-routes.

Once route planning is complete, the vessel is redeployed, now having the data previously collected and stored for rapid recall on disk. During this route survey operation, the vessel is guided along tracks, parallel to the intended Q-route, precisely locating and classifying all contacts of significance. The information from sonar is related to position and displayed on a video terminal, significant contacts are extracted and fed to the AIS for display on the command console. The minehunting sonar is also directly interfaced to the AIS during this operation for additional contact location. The AIS display therefore shows an overview of the Q-route with all contacts and their classifications, and all route data is logged on disk.

The final phase of the operation is conducted when a mining threat is suspected. The vessel is deployed with the AIS configured for a full minehunting role. Contact data previously collected is recalled and displayed on the command's ground stabilized display. The vessel follows the Q-route, investigating only those contacts not previously located, so a rapid clearance is accomplished and full shipping freedom restored.

Precise Navigation Buoy

The Precise Navigation Buoy (PNB) shown in Figure 6.3 is an example of a buoy that was designed to provide a precision navigation benchmark for minesweeping operations in depths up to 60 meters. Applications using visible, radar or radio beacons can be easily implemented, and the discus buoy is an excellent wave-rider instrument platform. The PNB is easily and automatically deployed from either surface craft or helicopter. It provides:

- Accurate sea-surface positional reference
- Absolute wave displacement reference—sea surface to bottom
- Relative wave displacement reference—sea surface to platform
- Constant line tension independent of wave, tide or current conditions

EDO Co. Navtrak V

An ultrashort baseline acoustic tracking system. It can simultaneously track up to five transponders or responders attached to vehicles or moored to the sea bed or subsea structures. The transducer design is less than three inches in diameter and allows tracking in the full 180-degree hemisphere below the vessel.

EDO Co. was awarded a 1986 contract by the British Royal Navy's Admiralty Research Establishment (ARE) for Navtrak V systems that will be installed on Royal Navy "Hunt" class minehunting ships and used to track PAP-104 MCM ROV systems. The Navtrak systems are being built by EDO Ltd. of Almondbury in Britain.

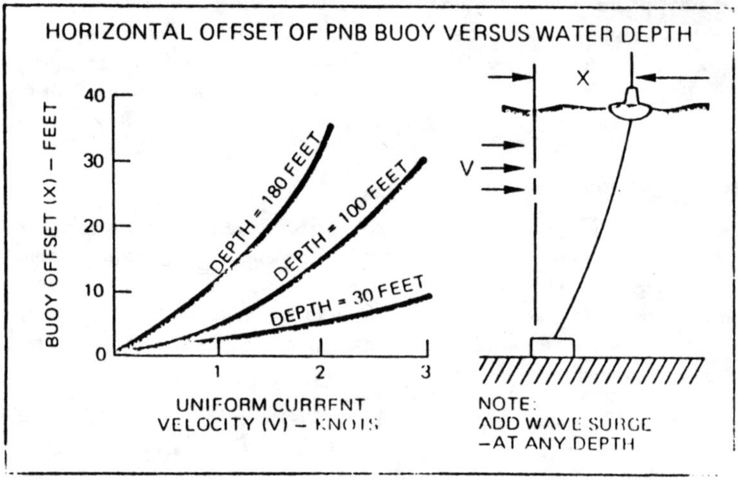

Figure 6.3. Precise Navigation Buoy (PNB) provides a Precision Navigation Benchmark for minesweeping operations.

EDO Co. has recently introduced a miniaturized version of the Navtrak V system. This is designed to operate as an integrated component in a shipboard navigation system, known as MicroNavtrak. It provides the same acoustic accuracies as Navtrak V but does not include a monitor to display the targets. Instead, the output from the processor goes into the shipboard computer in several choices of data format.

Quibit TRAC IVB

An integrated navigation system that has been used for trans-Atlantic cable route survey. It was installed and operated for Cable & Wireless (Marine) Ltd. onboard the

MV Cable Protector for route survey from Weston-super-Mare (near Bristol) in Great Britain to New York City in the U.S.A. The TRAC IVB was programmed to continue surveying on dead-reckoning in the event the Loran C equipment faded in mid-ocean. This navigation system could have application in mine-warfare operations. See Figure 6.4 for a view of the system operating configuration.

AQUANAV

This is an acoustic navigation system for real-time position plotting of an MCM vessel, ROV or towed vehicle. See Figure 6.5 for a view of the system components, and a plot of ship and ROV position.

The AQUANAV uses an eight-channel acoustic interrogator with RS-232 output to IBM-compatible portable computer, and Datasonics Autocal Transponders to perform automatic sea floor transponder array calibration. System components are available in different frequency bands to cover all range and depth requirements.

The manufacturer is Datasonics Co. of Cataumet, Massachusetts in the U.S.A.

NAUTIS-M

A Plessey Marine Ltd. display that gives a good operational picture of mine-classification maneuvers and mine-disposal operations. See Figure 6.6 for a picture of the plot.

UQN-5 SANS

Swimmer Area Navigation System. Designed and produced by Allen Osborne Associates. Used by Naval Special Warfare Group (NSWG) Seal Team divers to perform grid searches of the sea bottom in conditions of reduced visibility. See Figure 6.7 for

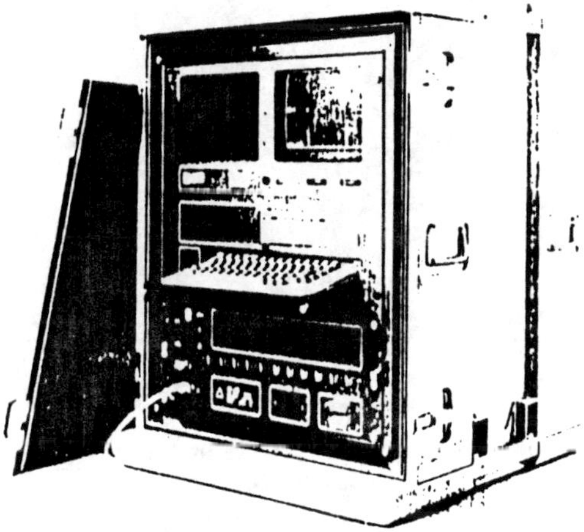

Figure 6.4. Quibit TRAC IVB Integrated Navigation System.

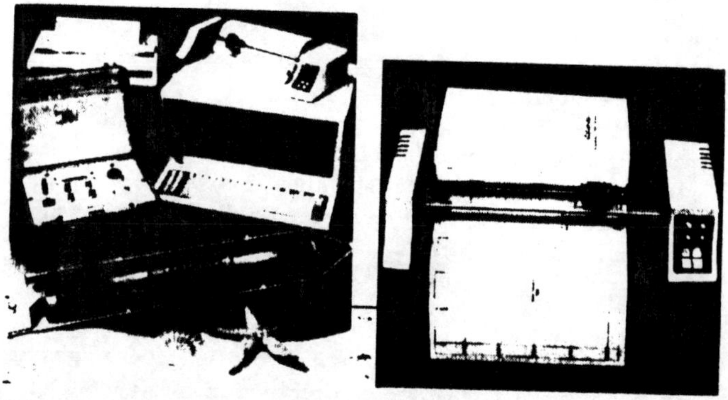

Figure 6.5. An AQUANAV Acoustic Navigation System for real-time positioning of vessel, ROV or towed vehicle.

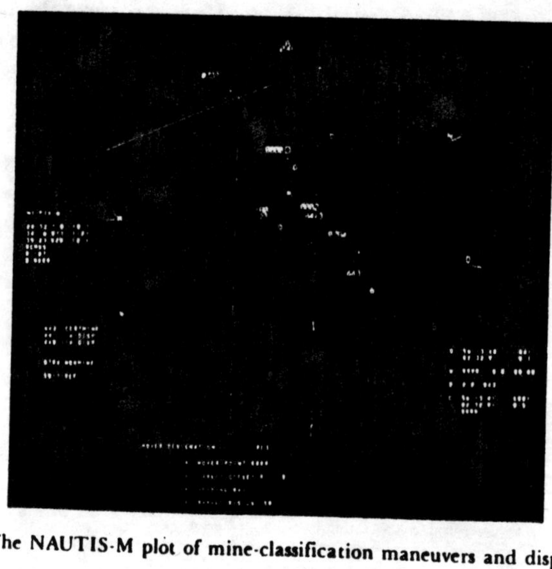

Figure 6.6. The NAUTIS-M plot of mine-classification maneuvers and disposal operations.

Figure 6.7. Components of the SANS (Swimmer Area Navigation System). Used by U.S. Navy Seal divers.

a view of the components. The system employs a diver-held receiver unit synchronized with two acoustic beacons placed at extreme corners of the desired grid pattern. The diver is given a real-time readout of position and heading and the ability to record the location of objects on the sea bed. At completion of the dive, the receiver is attached to a printer that produces a hard copy record.

V-COM

Underwater vehicle communications system. The V-COM is a tri-mode vehicle communications system used by U.S. NSWF on its Swimmer Delivery Vehicles (SDVs). Figure 6.8 shows the components. The electronics subsystems consist of an intercom, an underwater telephone and an HF radio, each housed in a pressure-tight aluminum canister. The intercom provides communications between the pilot, navigator and the crewmen of the SDV. The underwater telephone is compatible with standard U.S. Navy acoustic communications equipment and provides a link between the underwater vehicle and its support ship. The HF radio contains a modified man-carried unit (PRC-104)

Figure 6.8. V-COM underwater vehicle communication system.

and enables radio communications when the vehicle is surfaced. This system is also manufactured by Allen Osborne Associates of Westlake Village, California.

Raydist

A precise navigation system used on RH-53 and MH-53 U.S. Navy minesweeping helicopters. There has been considerable down-time with this equipment—reliability and maintenance problems.

A-Tracker

The U.S. Navy Air Systems Command in 1988 published a synopsis on development and testing of an underwater ultra-short-baseline navigation and tracking system called the A-Tracker to be used with the Airborne Mine Neutralization Set (A/A 25E-24) and various transponders. The system would be installed and operated from RH-53D and MH-53E helicopters. The equipment would be used to locate positions on the sea bottom that have been previously marked by transponders and to track remotely controlled underwater vehicles.

The A-Tracker will consist of an underwater transducer assembly that will contain:

- Variable-depth acoustic transducer
- Electronics
- Sensors
- Mechanical components
- Rack-mounted control/display console in the aircraft

Signals from the transducer assembly will be received, processed and the final result displayed on the console. Video data will display the relative positions of the various transponders being interrogated, indicating headings of the underwater vehicles being tracked. A tethered cable will provide transmitted electrical signals between the underwater transducer assembly and the helicopter console.

CHAPTER 7

Degaussing of Vessels and Vessel Construction Techniques

HISTORICAL BACKGROUND

When the German Navy deployed magnetic mines during the early part of World War II, there was a frantic effort on the part of the British to come up with a countermeasure. They put their best talent to work on developing methods to measure ships' magnetic fields and then devising means of countering each ship's magnetism. The U.S. Navy was attempting to get information on what the British were doing, and at first very little information could be obtained. The U.S. Navy was conducting experiments on the *U.S.S. Vincennes*, but at the same time the U.S. Naval Attaché in London was asked to obtain information on the countermeasures which the British were known to have developed over the previous months. Some information was received in May 1940. However, in March 1940 the unfinished liner *Queen Elizabeth* arrived in New York with a large degaussing girdle for all to see. A fairly complete description of the equipment was obtained from the *Queen Elizabeth* crew and from other British merchant vessels in American ports. In April 1940, the U.S. Naval Ordnance Laboratory received a piece of degaussing cable from the Commandant of the Third Naval District along with the startling information that more than 30,000 meters per day were being manufactured by U.S. companies for the British. Thereafter, degaussing coils similar to the British type were installed on U.S. Navy ships *U.S.S. Omaha*, *Barry* and *Goff* in June 1940. After this, the Chief of Naval Operations directed that coils be placed on all U.S. Navy vessels.

Definition

Degaussing is the process of eliminating, or rendering insignificant, a vessel's magnetic signature. Early methods of attaining this were necessarily makeshift and crude, but nevertheless effective. They involved the use of extensive lengths of cable and high electrical currents. Degaussing is still a primary mine countermeasure, especially for mine-sweeping vessels themselves.

Today, specially built mine-countermeasure vessels are constructed in such a manner that their magnetic signature is minimized from the outset by the use of non-magnetic materials. Wooden hulls have been superseded in most navies by glass-reinforced plastics or special non-magnetic steels. Engines are also built of non-magnetic metals, and all ferrous materials have been reduced to the minimum to reduce magnetic signatures. Some equipment, however, must unavoidably contain the offending high-signature materials and must therefore be degaussed. Electronics plays a major role in this process.

Ships can to a certain extent be protected against the magnetic mine, as was done in World War II. This is achieved by fitting coils carrying an electrical current around the hulls to produce an equal but opposite polarity magnetic field to the field already existing in the ship. This is degaussing. The ships go over a range which measures their initial magnetism, and from this it is decided what sort of coil protection is needed. They then go over the range again to see how effective it is. When a vessel goes into a shipyard for work, the magnetic field is likely to be changed, and the whole process may have to be repeated.

Diesel engines used on mine-warfare ships are normally fabricated of special non-magnetic stainless steel alloy to help reduce the possibility of detonating magnetic mines. Naturally, this adds to the cost of the engines and overall vessel cost.

Another method of reducing magnetism, particularly in small ships such as MCMVs is to build their hulls of a non-magnetic material such as wood, GRP (Glass-Reinforced Plastic), or low-magnetic steel. Wood is still used, for example in the U.S. Navy MCM-1 "Avenger" class of MCMV. The types of wood used include Douglas fir, white oak and Alaskan cedar. A plastic coating is used over the wood, however. Special beryllium-copper wood screws are used to attach the planks of wood to the frame members. See Figure 7.1 for a view of one of these wood screws that are used on the "Avenger" class construction in the Peterson yards at Sturgeon Bay, Wisconsin, U.S.A. These are non-magnetic, but are quite expensive.

GRP, of course, is non-magnetic and stronger than steel. However, as a mold has to be made and the GRP poured into it, this is not suitable for ships much larger than MCMVs, which average 500 to 700 tons.

In West Germany, the hull of the FGN's 10-ship series of Type 343 MCMVs are constructed from non-magnetic steel. Lemwerder yard of Abeking and Rasmussen in West Germany at Bremen (Lemwerder), Lurssen at Bremen (Vegesack) and Krogerwerft at Kiel are all building some of these vessels. See Figures 7.2.1 and 7.2.2 for views of a new minesweeper being constructed of non-magnetic austenitic steel.

The special steels of nonmagnetic type were first used in the type 205 German post WW II class submarines. The first two suffered from corrosion problems so severe that they had to be rebuilt with more conventional materials before a suitable non-magnetic steel was found.

In the case of the Type 343, stainless steel was selected after exhaustive tests in competition with GRP and wood, and no such problems are expected. However, the grade was chosen for its non-magnetic rather than its anti-corrosive properties. The potential differential between the hull and fittings (such as the cycloidal propellers) of a different grade have necessitated a complex system to ensure the elimination of electrolytic action.

Nonmagnetic steel (NM-steel) designated PN 18 S2 was successfully used by the Federal German Navy in the construction of the class 206 submarine. The fully welded steel construction is characterized by a thin-walled sheet metal structure having a plate thickness of 4 to 6 mm, as well as narrowly spaced struts.

Figure 7.1. Note the special non-magnetic beryllium-copper 13-cm wood screws used to construct U.S. Navy MCM-1 Avenger class minehunters.

Many factors must be considered in selecting the "ideal" material for construction of mine-warfare vessels. The competitors are:

- Wood
- GRP single skin
- GRP sandwich
- NM steel

No significant differences in construction costs can be found. They range within five percent. NM steel does cost slightly more, but this cost is compensated for by the advantages of good electromagnetic compatibility, grounding, storage of liquids, and ease of acoustic problem resolution.

Welding of the special non-magnetic steel requires a special skill apparently not available in all countries. One of the problems encountered during the manufacture is that the shrinkage is about three times as high as with standard shipbuilding steel when heat is applied, as in welding and cutting. Some special jigs and fixtures may have to be fabricated. Tools coming into contact with the metal must be made of stainless and non-magnetic materials so as not to contaminate the basic material. Maintenance or

Figure 7.2.1. The German Navy's new minesweeper *Ueberherrn* is built of non-magnetic austenitic steel. Shown here in final stages of completion.

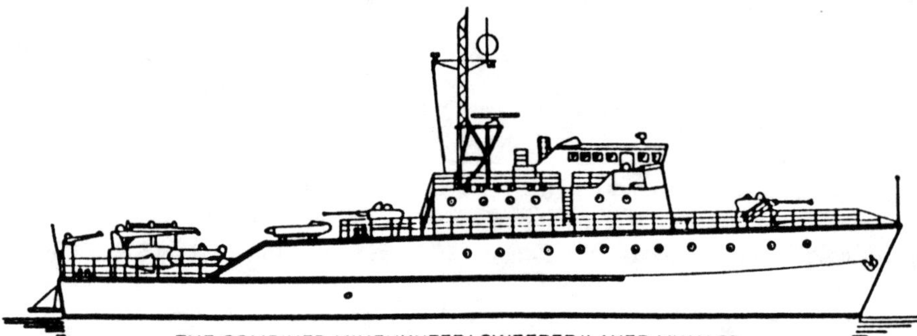

THE COMBINED MINEHUNTER/-SWEEPER/LAYER MW-V 50

Figure 7.2.2. The Type 343, MW-V 50 combined minehunting/sweeping/laying vessel built by Abeking & Rasmussen for the German Navy. Although the company can also build with wood and GRP, amagnetic steel was chosen after careful research.

modifications are very easily made in the NM steel vessel. A hull made of steel requires the least sound-related extra effort. Sagging in a sea state is minimal with the steel hull.

It is interesting to note that the U.S.S.R. uses steel hulls for some of its mine-warfare vessels. Some of these have been given or sold to other nations.

The Karlskrona shipyard of Sweden has designed the "Type M7O" minehunter for the Royal Swedish Navy. The ship's integrated design includes a totally non-magnetic hull made of alternate layers of glass-fiber and reinforced plastic. This "sandwich" type of construction does not require a mold and thus has certain advantages over those construction techniques that require molds. See Figure 7.3 for a cut-out view of the hull section.

The British Royal Navy and Vosper Thornycroft Ltd. have incorporated a new generation of developments to counteract the threat from intelligent mines by using improved detection and disposal techniques. The SRMH (Single-Role Minehunter) is made considerably less visible to sensing mines by adopting tighter control of magnetic and acoustic signature emissions and of EMC (Electromagnetic Compatibility). The "Sandown" class mine-countermeasure vessel was ordered by the Royal Navy in 1985 and the First of Class was launched in 1988. See Figure 7.4 for a view of the *H.M.S. Sandown*.

The "Hunt" and "Sandown" class MCMVs have been constructed entirely of glass-reinforced plastic. This material has been easy to fabricate, but a major drawback is that it is transparent to electromagnetic radiation. To correct this problem, it has been necessary to use a flame-sprayed metal process on the surfaces of selected compartments for sheilding, and extra screening and filtering for many items of equipment. If a poorly RF-shielded ship comes into the proximity of high-powered radars operating in a high-frequency band, severe interference can result aboard the poorly shielded vessel.

An acoustically noisy bow thruster was eliminated by the use of two Voith Schneider propulsion devices positioned aft. An automatic heading control using an autopilot is provided.

Various Degaussing Systems

AEG of Frankfurt, West Germany has developed a degaussing system for the active protection of magnetic ships, by reducing the ship's magnetic disturbance field to such

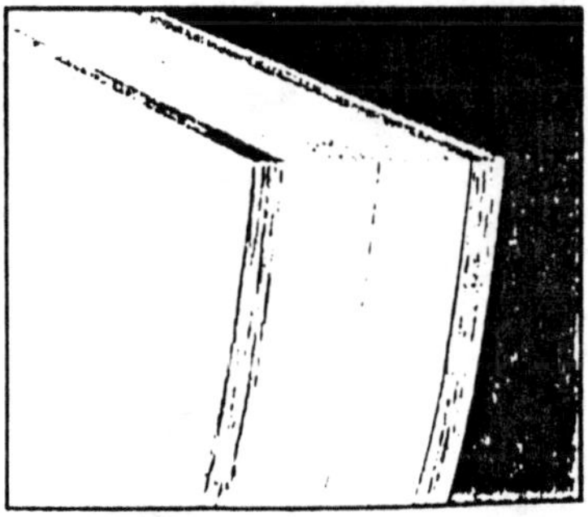

Figure 7.3. Hull Section of Swedish reinforced GRP "sandwich" construction.

Figure 7.4. Royal Navy *H.M.S. Sandown*—single-role minehunter. Uses GRP construction.

an extent that activation of magnetic mines is minimized. This is achieved by means of a magnetic field that is opposed to the disturbance field. A coil system that is fed by a degaussing power plant is installed in the ship. AEG also provides deperming facilities. See Fig. 7.5 for typical coil arrangement.

Figure 7.6 shows details of a degaussing system designed by the Italian firm IFEN S.p.A. of Ceparana, La Spezia. The various coils are shown on a typical minehunter.

The Thomson-CSF IMAIP-3 Degaussing System was selected for the Tripartite minehunting vessels of NATO. The French firm developed the IMAIP-3 system at its Brest, France plant, for manufacture in Belgium under the Tripartite agreement. The system consists of a three-axis magnetometer and a single electronics cabinet containing an adjustment subsystem and amplifiers. The approach used offers a high degree of modularity and consequent flexibility, permitting the system to be adapted to any type of minehunter or minesweeper. The same technology may also be used for steel vessels, but the degaussing process for these is from four to ten times greater than for non-magnetic ships, depending upon the number of circuits and amplifiers and their power. The power required in the case of the IMAIP 3 minehunters is 100 watts but

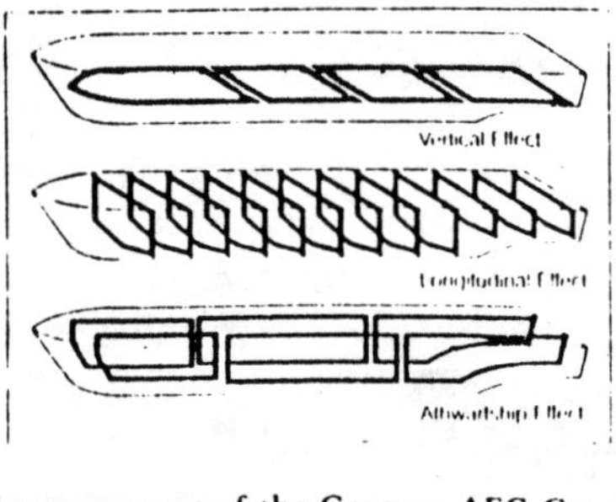

Figure 7.5. Coil arrangement of the German AEG Co. degaussing system.

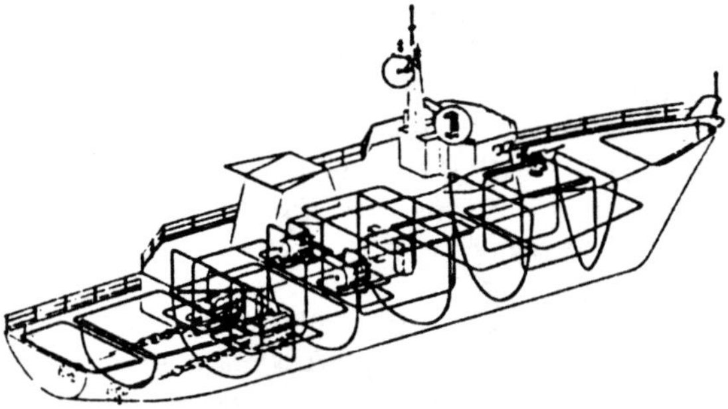

Figure 7.6. Degaussing system designed for Italian Navy by IFEN S.p.A. of Ceparana, La Spezia.

may be as much as 30 kilowatts in the case of a frigate. The use of electronic-control technology has eliminated the requirement for all rotary machinery and its attendant problems of maintenance, additional weight and noise.

Thomson-CSF has also developed magnetic sensors for the purpose of making magnetic measurements in conjunction with degaussing ranges both onshore and at sea. These facilitate the gauging of the actual degree of degaussing required. Advances have been made in the application of computers and in the design of sensors applicable to the type of measurements required. These sensors take the form of a three-axis type, with a cardan suspension for taking measurements while at sea, and mini-magnetometers for shore-based measurements. The concept permits degaussing systems to be automatically controlled.

Marconi Command and Control Systems Ltd. of Surrey in Great Britain was to have supplied its MDG 1702 magnetometer degaussing system to the U.S. Navy for the "Cardinal" class minesweeper/hunter, which has been cancelled. However, Marconi does supply the 1701 type system for use on the U.S. Navy "Avenger" class MCMV. Because the MDG 1700 series is a lightweight modular system, it has been possible to use the same equipment modules for several classes of vessels. The 1700 series is also used on Royal Navy single-role minehunters and by the Royal Australian Navy on its minehunting catamarans.

Marconi also builds an SDG 1800 series of gyro degaussing equipment. This is for steel ships and submarines. These degaussing systems require little operating attention, a minimum of servicing and, because of commonality of many of the components, require very little logistic support. This is classified as a "User Friendly System." Marconi uses either a gyro-controlled or a magnetometer-controlled system.

In the gyro system, a computer receives the ship's heading, roll, pitch, magnetic variation and zone information and supplies three analog signals. These signals are fed to power amplifiers that provide the coil currents. In this system, the operator is required to set into the equipment the value of H and Z, the earth's magnetic field components, in accordance with the magnetic zone in which the vessel is operating and the variation at that point. The equipment then varies the coil currents smoothly, taking account of the vessel's heading, pitch and roll. Provision is also made for mixing in a steady current to counteract permanent fields.

In the magnetometer system, the analog signals are produced proportional to the actual ambient field, rather than using computed ambient field. This is accomplished by sensing the field above the vessel in three axes by means of fluxgates and deriving the analog voltages from these.

For this type of degaussing, the basic control system consists of a masthead-mounted three-axis magnetometer, a control unit or cubicle, an appropriate number of power output stages depending on the size of the vessel and the required degree of degaussing, and a bridge control unit. The control unit interacts with the fluxgates to provide DC signal levels directly proportional to the field strengths in each of the three axes. Current is then passed to the degaussing coils as required to cancel out the ship-generated fields along these main axes and maintain the vessel's magnetic invisibility, assisted by output signals dedicated to supplying compensation specifically for eddy current fields.

Thorn EMI Ltd., Naval Systems Division in Great Britain has developed a new low-cost, portable degaussing system known as MS90, which should cut the cost from that of conventional degaussing ranges by over 50 percent. In addition, the MS90 will also cut operational costs and demands on the ship's sailing time by taking the range to the ship, and by completing the analysis of the ship's magnetic signature in about 90 minutes, compared with the two or three days required for conventional ranges. The MS90 uses only two self-surveying sensors, compared with the 12 to 18 sensors required in most previous systems. The twin sensor modules are packaged in a specially designed housing suitable for permanent immersion down to 50 meters. See Figure 7.7 for the MS90 system. For a transportable system, the data from the sensors can be sent to the computer system for processing by radio link from a telemetry buoy up to 5 km off-shore. The signatures can be displayed and printed for analysis and recording in exactly the same manner and with as much data as at present. The entire system can be housed in a standard 9-meter ISO container, which also houses the power supply generator.

NAVSEA Systems Command of the U.S. Navy has established magnetic criteria that ships, particularly minesweepers, must meet. Thus, all ships are tested at degaussing ranges and calibrations conducted if the criteria are exceeded. Magnetic silencing facilities (MSF) are also required for deperming and flash deperming of ships when the need arises. Some facilities incorporate special equipment for testing a minesweeper's stray fields and eddy currents.

Since 1980, the emphasis by CNO on magnetic silencing has increased. As a result, the use of degaussing services has increased. In order that facilities might keep pace with the increased demand, the magnetic silencing facility (MSF) program was initiated to overcome logistic problems created by outdated data acquisition and analysis at the

Figure 7.7. Deploying the portable telemetry buoy for the Thorn EMI Electronics MS90 Degaussing Range.

MSFs. The program provided for a complete upgrading of the existing facilities and the creation of two others. The upgrading was scheduled for completion in 1989.

Ship Maneuverability

In order to be fully effective, minesweeping vessels must have great maneuverability. One of the best methods of achieving this is the Voith Schneider propeller. This consists of pivoted propeller blades projecting from a circular casing flush with the ship's bottom, rotating about a vertical axis at constant speed. The blade shafts are fitted with links connecting them to a steering center. By moving the steering center from the center of the circle, the blades are made to describe an oscillating movement about their axes and a thrust is produced at 90 degrees to the direction in which the steering center is moved. By this means, the ship can be made to perform any desired maneuver, including turning on the spot—and even moving sideways, if two Schneider propellers are installed. The thrust can be adjusted steplessly in magnitude and direction, and the prime mover runs at constant speed without reversal being necessary. This saves time when performing fast maneuvers.

See Figure 7.8 for a view of the Voith-Schneider propellers.

The only protection against the acoustic mine is to make ships more silent. This is a complicated matter, very expensive, and the best minds in the world have been working on it for years without a great deal of success, except perhaps in submarines.

Special so-called noiseless propellers are available and have had a certain amount of success, but a ship makes a lot of other noises. Recently, the U.S. Navy was greatly disturbed by the Soviet purchase of certain Japanese and Norwegian computers and mechanical equipment that has enabled the Soviet submarine makers to build quiet

Figure 7.8. Voith-Schneider propellers for minehunters. This design gives excellent maneuverability to the ships.

propellers and thus make the Soviet submarines more difficult to detect with acoustic sensors. Besides the ship engines, any form of auxiliary machinery may also be noisy enough to trigger an acoustic mine. The mere passage of the ship through the water creates flow noise from drops of water falling off the hull. Diesel engines are particularly noisy.

Mine "Guinea Pigs"

The German Navy in World War I used mine-barrage runners to guide warships through especially dangerous areas. These mine-barrage runners were deep-draft commercial vessels on the order of 4,000 to 6,000 GRT whose holds were filled with wooden beams, empty drums and sand. This permitted them to withstand one or two mine hits. Out of 16 such vessels used, eight of them perished in mine barrages.

The British used such vessels to help in clearing mines in World War II. This is still considered by many to be a practical method of clearing pressure-type mines.

In 1954, the U.S. Navy conducted tests of five transports converted into "Destructors" to be used against pressure mines. The most successful turned out to be a ship of the "Liberty" class. The MSS-1 special minesweeper was therefore created. In 1968, after thorough tests, it was put into reserve.

Various other proposals have been made for the use of such vessels in conducting mine-clearance operations. One idea included the use of ping-pong balls stored in the holds.

CHAPTER 8

Typical Mine Warfare Operations

MINELAYING

The platforms generally used to lay mines are submarines, surface ships and aircraft. As demonstrated during the Korean War and recently in the Gulf of Suez and in the Persian Gulf, surface ships of many types can be used, including RO/RO with ramps for easy deployment.

Figure 8.1 shows some of these methods of deployment.

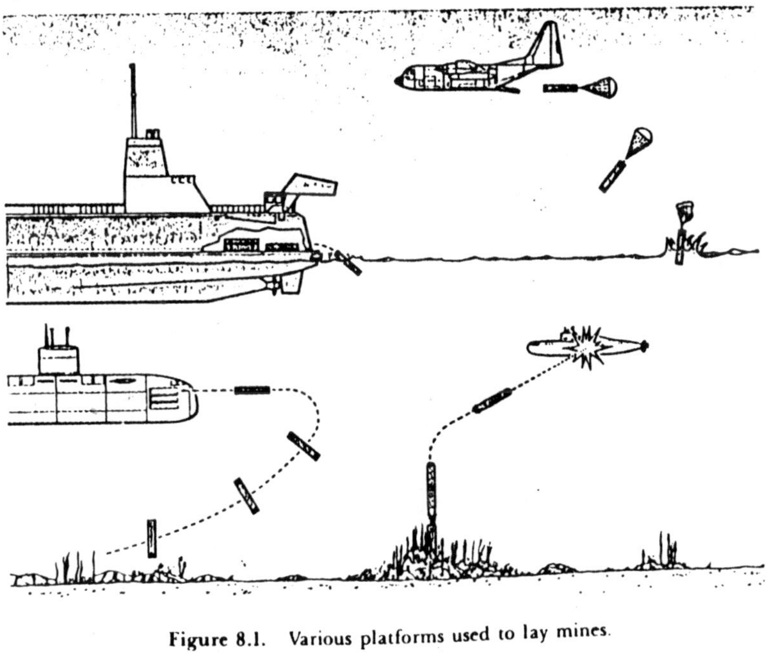

Figure 8.1. Various platforms used to lay mines.

Considerations in Minefield-Laying Procedure

Planning a Minefield—

Steps to be taken:

1. Performance objective.
2. Where and what targets? Primary and secondary.
3. Purpose—harass, close port, cut transport, etc.
4. What delivery assets does Fleet plan to use?
5. Where are mines coming from? MOMAG site, carrier, etc.
6. MOE—What Measure of Effectiveness is desired?
7. Intelligence.
8. How long will build-up of mines take?
9. How long is field to last?
10. Will operational command accept lower threat?
11. Authorize replenishment of field?

Analysis

1. Intelligence collecting.
2. Environmental conditions.
3. Choose location and determine mine settings.
4. Murder board—evaluate. Critically examine planning for errors.
5. MFPF—Minefield Planning Form Folder.

Minefield Planning

The following are some considerations and methods to be taken into account when planning minefields. NATO countries rely on computer calculations on how many vessels are needed to clear a given area of mines or to plant mines. The computer will come up with a confidence factor based on the data input.

OP-2637 Vol I —Characteristics of U.S. Mines
Vol II—Actuation & Damage Data

Damage Zones: Kill
M Mission Abort Damage-Loss of weapon delivery capability
C Communications & Nav impaired
R Repairs onboard possible

Sensitivity setting on Mines
Ship Count: Number of ships to go by before mine goes off
Soviet submarines are well degaussed
Speed of target
Acoustic signature

Reliability of MK (enter type mine) = ______%

Probability Curves drawn—by computer

Damage Width: Mines go off and do prescribed damage to target

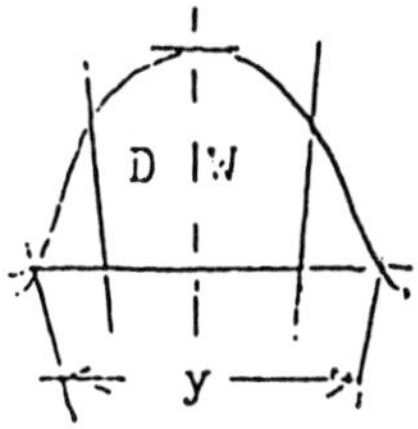

$$\frac{W_d}{\text{Useful fire ratio}}$$

Simple initial threat—threat to 1st transistor entering field
(50% is minimum)

Random Minefield in channel $T = 1 - (1 - \frac{W_d}{B})^m$

m = # mines in channel
W = Damage Width
B = Width of channel in ft.

Arca Mine $T = 1 - (1 - \frac{W_{dx}}{A})^m$

x = shortest distance in ft ship can sneak through minefield

$$\text{Total THR} = 1 - \left[(1 - T_a^{.75})(1 - T_b^{.65})\right]$$

A portable hand calculator can be used to make the calculations.
Ships, stations, or schools may use computers.

Casualty Distribution:
Another measure of effectiveness
How many casualties = 2 out of 10 during life of minefield
90% Confidence level

MOE—Measure of Effectiveness—sustained threat
Example: 30% over life of field or ______% at end of life of field

NWP-26A = Mine Operations
NWP-26-1 Mine Planning

World War II U.S. Army Air Corps Mine Operations

The XXI Bomber Command B-29s laid mines in Japanese waters during the latter part of World War II. The U.S. Navy (Admiral Nimitz) worked with the U.S. Air Corps (General LeMay) in the planning and execution of this attempt at massive mining of Japanese home waters.

In order for an actual position of the mines to be recorded, the APQ-13 radar on the bomber aircraft was used to determine from geographical scope photos and known mine ballistics, including the effect of wind drift, where the mines actually were planted.

The mines deployed by these bombers around Japan caused more ships to be sunk than were sunk by U.S. Navy submarines with their torpedoes.

Figure 8.2 shows a typical operational scenario of a weapon such as an encapsulated torpedo-type mine. In the case of a Soviet type of such a weapon/mine, the torpedo would

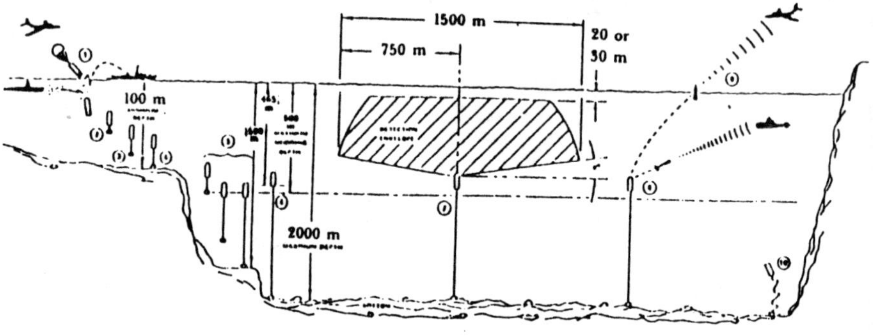

Figure 8.2. Operational sequence for mine such as "CAPTOR" Mk-60.

Sequence:
1. Launching
2. Case/anchor separation
3. Initial payout of mooring line
4. Shallow water mooring
5. Payout of mooring line. Case held at 500 mtrs in bottom depths to 1500 mtrs.
6. Deep water mooring depth 400 mtrs in bottom depths 1500 to 2000 mtrs
7. Detection system actuation
8. Submarine detection and firing
9. Buoy release and data transmission
10. Scuttling
    End of preset life
    Above 90 meters
    Battery power failure

Note: All measurements in meters.

be replaced by a missile-type device that would react much quicker (because of its greater speed) to intercept the target vessel.

MINE COUNTERMEASURES

General

In the broadest sense, naval mine countermeasures includes all possible means of preventing sea mines from destroying ships and submarines. This involves the following:

- Preventing the enemy from laying mines in the first place.
  This can even mean attacking his industries that produce mines, his operating bases, his platforms that can lay mines and interdicting his lines of communication.
- Preventing our own ships from actuating mines.
  This includes methods such as degaussing, building quiet ships and operating at low speeds with shallow-draft vessels to prevent detonating pressure-actuated mines.
- Finding mines once they have been laid.
  This includes laying out "Q" routes and route survey, even before mines have been laid, to identify all mine-like objects that could hamper mine locating and clearing work at a later time.
- Sweeping the mines.
- Destroying or neutralizing the mines.

Operational Factors

To clear an area of a mine threat using sonar, a long operation involving the investigation of bottom contacts of mine-like objects (MLO) has to be undertaken. The duration of this type of operation in time of conflict is far too great when the rapid deployment of shipping is required, so the concept of "route survey" has been introduced. A suitably equipped vessel is employed to collect accurate hydrographic data, initially along strategic routes (Q-routes) and then in areas of less operational importance.

This data will include all sonar contacts in the area with appropriate classifications and may include some shallow sub-bottom profiling. The information will be updated at a frequency according to the nature of the sea bed and operational constraints. In time of conflict, the data collected is recalled and the minehunting vessel redeployed along the Q-routes, investigating only those contacts that have not previously been encountered. Thus, a rapid detection and, if necessary, clearance is achieved.

This route-survey concept lends itself to a combined-role minehunting vessel. In peacetime it can be deployed, collecting valuable hydrographic data that can be used for both military and civilian applications, while in time of conflict a small adjustment will enable a full minehunting role to be assumed. This concept also provides excellent facilities for operators, as the same equipment is used for the various tasks.

For the combined-role vessel to be capable of giving a credible performance, the sensors and Action Information System (AIS) must be capable of fulfilling all the requirements of dedicated minehunting and hydrographic survey. The primary sensors will include:

- Minehunting and side-scan sonars
- Precision survey echo sounders
- Remotely operated vehicle (ROV)

These sensors interface with an AIS that will provide operational displays and full data recording. Radio positioning systems, when interfaced with the AIS, will provide the accurate positioning required for vessel guidance.

Channel Conditioning and Route Survey

One function of channel conditioning is to identify the density and types of mines and mine-like objects in these channels. If a military commander decides not to clear an area of mines, Q-routes will be selected to avoid any area where mines are located. Q-routes are those channels leading from a port to the open sea.

Other nations have a variation of the U.S. approach to mine warfare. Many of the policies of these nations reflect an attitude in which the only purpose of mine warfare is to protect their own ports and coastlines. Also, some nations would like to rely upon their allies to perform the major functions of minehunting and clearance. However, it is believed that most are now realizing the value of developing a peacetime activity for channel conditioning.

Channel Survey involves searching for and charting all mine-like objects in channel and harbor approaches, prior to a threat of deliberate mining. A comparison of past surveys with more recent ones would reveal possible mines and their positions.

Channel conditioning is a process that involves the systematic removal of all mine-like items, leaving a "sterile" channel. This reduces the number of contacts that would require investigation during a suspected mining incident. Channel conditioning should be actively pursued to eliminate any mine-like objects such as oil drums, refrigerators, cars, etc. If a true threat then develops, the channel conditioning would speed up the investigation of all suspect items. Channel conditioning is sometimes referred to as sea bed surveying.

Route survey lends itself to a combined-role minehunting vessel. In peacetime it can be deployed, collecting valuable hydrographic data that can be used for both military and civilian applications. In time of conflict a small adjustment will enable a full minehunting role to be assumed. This concept also provides excellent training facilities for operators, because the same equipment is used for the various tasks. It is believed that some of the East Block Soviet countries use this approach.

The "Q-Route"

Channel surveys and channel conditioning should not be confused with the current "Q-system." The Q-system or Q-route originated with the British, and was adopted by the U.S. Navy during World War II and modified to meet the requirements of both countries. The Q-route normally begins at a harbor entrance buoy and extends out to sea, ending at a designated depth where the mine threat is no longer significant. These routes are not as critical as choke points and are flexible to operational planning. Channel surveys would cover the deep-water passages from the channel entrance buoy to the port.

The Q-system serves as the classified portion of the navigational information systems for allied nations. The system is set up in such a way that naval ships that require full information on minefields will receive it. The remainder of naval ships will receive only such information as may be necessary for the proper execution of their missions. Merchant ships are provided with sufficient information to enable them to adhere to the searched channels, or where such do not exist, to stay with routes that will take them clear of all known dangers. AHP 1, Allied Navigational Information in Time of War, provides Q-system guidance.

Needless to say, navigation accuracy is of utmost importance to any of these mine-warfare operations. A number of times during past wars, even friendly vessels were damaged by our own mines, either because of poor navigation or a captain who "did not get the word."

Typical Port Breakout Considerations

Note: Considering a typical Fleet Command – Pacific
Ship Count – assume maximum of 21

Depart Port

Precise navigation – Raydist Stations
  4 ports East Coast, 5 ports West Coast

Pre-set routes out of ports

Channel Conditions – Route Survey

Navy assets

CINCPACFLT
  Main ports – use U.S.N. Reservists

RHDCC—Reserve Harbor Defense Coordinating Components
Planning—use Coast Guard and EOD units
MIUW Teams—Mine In-shore Underwater Warfare
Use craft of opportunity
Coverage out to 25 miles—radar range

Considerations

1. Command & Control—Reserve Naval Base Commander
2. Surface & Sub-Surface surveillance
3. Surveillance and Zone Patrol
4. MCM Plan
5. Harbor Traffic Control
6. Ship Inspection
7. Harbor Clearance
8. EOD
   Use ELC—Explosive Line Charges to blow out of harbor

Magnetic Sweeps

In principle, the magnetic sweep creates a magnetic field astern of the minesweeper as similar as possible to that created by a ship passing through the water. In its simplest form, it consists of towing two long cables astern of the sweeper. One cable is longer than the other, the difference in the two lengths representing roughly the length of a ship. At the end of each cable is an electrode, and a high pulsating current is passed down the cables. The two electrodes set up a magnetic field between them which resembles that of a ship and should explode any mines over which they pass.

Mines can be set to count, therefore it is often necessary to sweep an area a number of times. In general, however, once a possible minefield has been identified, it is customary first to carry out a clearance wire sweep. This is followed by acoustic and magnetic sweeps. If no mines are exploded after two or three sweeps, it is then necessary to revert to minehunting. Since this is such a time-consuming process, probably only a narrow channel would at first be cleared. It is also necessary to establish the limits of the minefield. Computers can be used to calculate how many sweeps should be made and how many vessels are needed to clear a given area. It is also possible for the computer to come up with a percentage confidence factor, based on the number and types of searches made.

British Royal Navy Operations

Figures 8.3 and 8.4 show magnetic sweeping equipment and mine-disposal weapons being operated from Royal Navy mine vessels. Figure 8.5 shows a Royal Navy minehunter with the new Plessey sonar for detecting mines, and the ARMS (Advanced Remote Minehunting System) ROV which classifies, identifies and can neutralize mines. Figures 8.6 and 8.7 explain the Royal Navy Sea Serpent Minesweeping concept.

French Navy Operations

The mine-warfare capabilities of the French Navy are well known and internationally recognized. Its minehunters have been requested to clear foreign and international waters

Figure 8.3. British Royal Navy coastal minesweeper *H.M.S. Glasserton* preparing TAG and Oropesa sweeping equipment.

Figure 8.4. Crewmen on Royal Navy minehunter lowering a disposal weapon used to blow up mines.

Figure 8.5. Above is a Royal Navy minehunter using the 2093 Type sonar to locate mines. Below, the ARMS (Advanced Remote Minehunting System) using an ROV to classify, identify and neutralize a mine.

British Royal Navy "Sea Serpent"

Cyclic Sweep

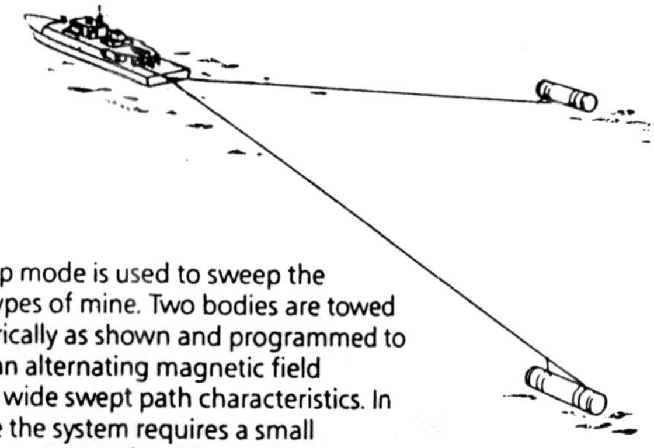

This sweep mode is used to sweep the simpler types of mine. Two bodies are towed asymmetrically as shown and programmed to produce an alternating magnetic field providing wide swept path characteristics. In this mode the system requires a small electrical power supply which could be provided by a portable generator. Three pairs of bodies may be towed in line astern.

Figure 8.6. **Sea Serpent mine-countermeasure device being operated in Cyclic Sweep Mode.**

Magnetic

This mode is used to sweep the more intelligent type of mine. 6 bodies are towed in line astern. By selecting appropriate signature levels for each body and suitable distances between bodies, the system can simulate the magnetic characteristics of particular vessels. Once the correct magnetic states have been set no electrical power is required until a further change is made.

Acoustic

The inclusion of other influence sweeps (e.g. Acoustic) with Sea Serpent in the magnetic simulation mode produces a High Performance Multi Influence sweep capable of eliminating a large proportion of intelligent mines.

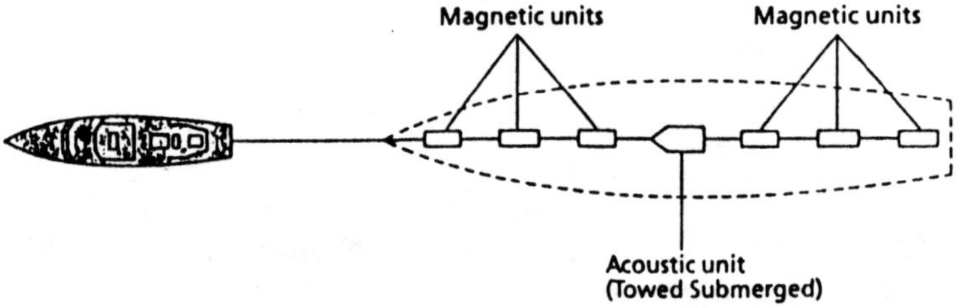

Figure 8.7. **Operation of the "Sea Serpent" in the Magnetic (left) and Acoustic Mode (right).**

on several occasions. In coordination with the logistic support vessel *Loire*, French "Circe"class minehunters participated in mine-clearance operations at the Suez Canal in 1974, 1975, and 1978 at the request of the Egyptian Navy. A division of minesweepers and minehunters was stationed in Djibouti in Eastern Africa from October 1980 to November 1982. They were on standby in case the Strait of Hormuz had been mined. Four French minehunters were again sent to the Red Sea with two support ships in August and September 1984. Two Dompaire-class hunters searched the Red Sea off Jidda and two Tripartite-class CMTs cleaned up their assigned sector in the Jubal Strait. Ten mines left over from the Yom Kippur War were detected and exploded at depths of 90 meters.

In 1987, three French minehunters were dispatched to the Gulf of Oman, after merchant ships were damaged by mine explosions.

At present, the French MCMs are limited to a depth of 80 meters in their operational depth. About 100 miles from the French Coast, the water gradually deepens to 300 meters, then the depths abruptly increase to more than 2000 meters. Since modern mines can be laid at greater depths, the French have adopted a five-year plan to include the production of ten new oceangoing minesweepers that will extend the French minehunting capability to a depth of 300 meters. These new vessels are called BAMOs (Batiments Antimines Oceaniques) or oceangraphic minesweepers. See Figure 8.8 for a diagram of how the French visualize the operational scenario for the new BAMO vessels.

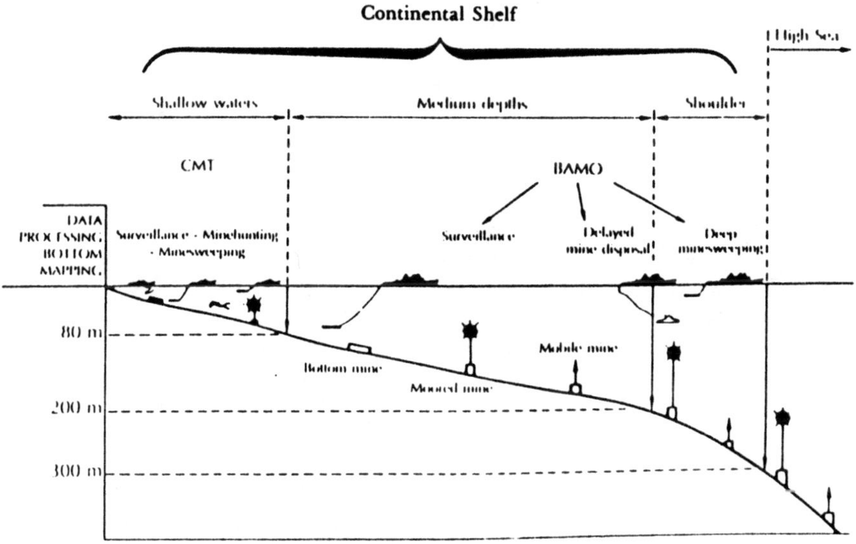

Figure 8.8. French Navy operational scenario for deployment of their BAMO oceanographic minesweepers.

U.S. Navy Mine-Neutralization System Operation

A Typical Mission

The mission begins with a mine-countermeasures ship proceeding along a section of water to clear an area of mines. The shipboard sonar detects a mine-like object suspended above the ocean floor or perhaps lying on the bottom. The MNS (Mine-Neutralization System) is then called into action.

On the ship's deck, the crew uses the VHS (Vehicle-Handling System) with its booms to lift the MNS ROV and lowers it into the water. Once it is safely beneath the surface and protected from wave action, the operator releases the vehicle with a burst of compressed air. The MNS vehicle is now deployed, connected to the ship by an umbilical cable that supplies electrical power and links it to the shipboard monitors and control consoles.

Aboard the ship, operators direct and track the ROV's underwater movements. Using range, bearing and depth data from the ship's minehunting sonar such as the SQQ-30 or the later SQQ-32, and the Honeywell RS/900 acoustic position indicator, the vehicle is moved toward the mine-like object. As it approaches, the vehicle's high-resolution, multimode sonar picks up the target. The operator maneuvers the vehicle in closer and focuses the forward TV camera to confirm that the target is a mine. He then guides the vehicle to the mine's mooring and clamps one of the two cutting arms onto the cable. The vehicle is then moved away, leaving the cutter attached to the cable. The MNS is then steered back to the ship for recovery.

From the control console, the operator sends an acoustic signal which fires a charge in the cutter. The explosive cutter severs the mooring cable and the mine floats to the surface, where it can be destroyed by gunfire.

The MNS then continues after the ship discovers another mine-like object, but this time it is lying on the sea floor. The MNS vehicle is sent out again to identify and classify the object, guided to the area by shipboard sonar until its own sonar and TV can find the target and verify that it is a mine. Because bottom mines are not moored and do not float to the surface, the MNS neutralizes them with a bomb-like charge. This charge must be dropped close enough to the mine to destroy it. However, it cannot be dropped too close or the mine may explode upon contact, destroying the MNS vehicle. As the operators delicately guide the vehicle toward the target, a computer-generated aiming ellipse on the control console is superimposed on the video image. The ellipse generates a target-to-drop zone distance, which changes as the vehicle moves closer and hovers over the target. When the ellipse is tangent to the mine, the vehicle is in the correct spot.

A charge carried by the MNS vehicle is released and drops to the bottom next to the mine. The vehicle is then steered back to the mother ship and is recovered. From the control console, the operator sends an acoustic signal which detonates the explosive, thereby destroying the mine.

The MNS vehicle emits low acoustic and magnetic signatures, enabling it to do its job undetected by mines. It has twin hydraulic longitudinal thrusters, along with vertical and horizontal thrusters, that permit maneuvers close up, even in moderate currents. A camera is housed in a bubble on the top, aft area, and is used during launch and recovery to ensure that the vehicle is properly attached to the handling system. This camera also provides the operators with a good view of the umbilical cable, allowing them to correct any tangling or twisting.

The vehicle carries two cable cutters for moored mines and a destruction charge for bottom mines. Once these have been used, the crew must recover, rearm and relaunch the vehicle for the next mission. A fiberglass frame and skin-housing sections of syntactic foam give the MNS positive buoyancy. In an emergency, such as cable entanglement, the operator releases a lead ballast weight from the vehicle and its umbilical is automatically severed. The MNS then surfaces and an emergency pinger and flasher are activated so it can be located and recovered.

On the ship, two MNS operators work side-by-side at the control and monitor consoles. The MNS location is determined by the RS/900 acoustic position indicator. A hydrophone picks up the acoustic signals from the vehicle's pinger and transmits them to the monitor console.

The control console operator controls the vehicle with a three-axis joystick. The operator at the monitor console studies the sonar display for target acquisition. Both operators are backed up by a constant and reliable flow of telemetry and computer software. These keep them advised of the vehicle's navigation track, altitude, depth and course.

The MNS vehicle-handling system enables a single operator to deploy and recover the vehicle. In addition to its three booms and capture saddle, the VHS includes a platform for vehicle stowage, maintenance and ordnance loading. The umbilical cable serves as the link between the MNS vehicle and the ship. It gives the vehicle good endurance, in contrast to other MCM ROV vehicles that are powered by batteries. The MNS umbilical-cable-handling system automatically controls cable tension, isolating the vehicle and the cable from the effects of the ship's motion. It also keeps the cable in good operating condition.

Wire Sweeps Equipment

Wire Sweeps (Oropesa). A method of dragging wires through the water so that they snag the mine's tethering cable. With the aid of a wire cutter embodied in the sweep, the tethering cable is cut and the mine bobs to the surface, where it is detonated or sunk by small arms fire. The sweep is kept opened out on the towing ship's quarter by means of kites and otters. Two sweeps can be fitted, one on each side of the ship, which can stream both sweeps at once or on one side only. Normally the single sweep is used, each ship keeping behind the ship ahead's sweep wire, so that each vessel has some slight protection, with the exception of the leading ship.

BAJ Ltd. of U.K. WS Mk 0 sweep includes a system which continuously monitors the tension in the sweep wire.

Acoustic Sweep

There are various types of acoustic sweeps. They are usually carried in some form of towed body in which a noise maker is placed. These can vary from simple devices in which a hammer strikes a gong to much more sophisticated equipment that simulates noises, as closely as possible, actually made by ships in the water.

Noise is made up of numerous different frequencies, and it is possible to simulate those most commonly found. The most sophisticated acoustic sweep yet produced is the British Aerospace Osborn Sweep, which is capable of emitting noise at a number of frequencies. It consists of an acoustic generator, towed well astern of the ship and kept just below the surface of the water by floats. Also towed is an acoustic monitor that feeds back to the ship the noises made by the generator. The noises can then be adjusted to suit the conditions and to represent different types of ships.

Typical Equipment for Conventional Minesweeping

The Kabelwerke Ehlers firm of Hamburg, West Germany produces five types of equipment for conventional minesweeping. These are as follows:

1. *Electrode units*

Comprised of two parallel cables of differing lengths (about 250 meters and 500 meters) with copper electrodes at their ends. The alteration in the magnetic field produced by the current of 900 to 1500 amps simulates that produced by a steel ship and thus explodes magnetic mines.

2. *Closed circuit devices*

Consists of a closed-circuit cable loop about 500 meters long by 60 meters wide that has the same effect on magnetic mines as the electrode units, with the advantage of a relatively wide sweeping range.

3. *Solenoid units*

A streamlined solenoid towed on a 300-meter-long cable towed behind the minesweeper to form a magnetic field that detonates the mines. It is highly maneuverable and requires only about 100 amps of current.

4. *Acoustic devices*

A device that produces acoustic (usually knocking) sounds. It is towed on a 300- to 500-meter cable. It covers a wide range of frequencies and simulates ship's engine noises that detonate acoustic mines.

5. *Wire sweeps*

Consists of two cables up to 650 meters long that are separated to cover a width of up to 200 meters. As soon as the mine-tethering cables are touched, explosive charges are detonated to cut them. This allows the mines to rise to the surface, where they can be destroyed by gunfire.

The Trawl Sweep Net—A Mine-Countermeasure Device

During World War II, trawl sweep nets were used for disposing of bottom mines. This system still has some merit for use in solving the mine-countermeasures problem. It is a cheap, effective, simple mine countermeasure.

The Office of Navy Research (ONR) and Woods Hole Oceanographic Institute conducted tests from 1954-57 using a 30-meter by 1-meter nylon mesh net. This net was successful in sweeping against bottom mines in relatively smooth bottom areas. Mk-36 and Mk-25 mines were used as targets.

The Naval Coastal Systems Center in 1964 performed similar tests on three different net designs. Although the results showed some promise, not much was done until the U.S. Navy Mine Warfare Command at Charleston, South Carolina prepared a plan of action to develop a trawl sweep. In 1984-85, with funding from the Naval Science Advisory Program and technical guidance from the Naval Coastal Systems Center at Panama City, Florida, the development of a single-vessel trawl net sweep was begun.

The prototype net consisted of six 30-meter panels and two 15-meter panels that could be interchanged to increase or decrease the size of the net. During tests on the NOAA research vessel *Chapman*, the net engaged all of the ten target mines. The speed of operation was about 4 knots.

Later tests in April 1986 by the Mine Warfare Command onboard the research vessel *Oregon*, a typical coastal trawler, were also successful. However, when in use against real mines, it should be kept in mind that some nets will be lost due to mine explosions. But they can be procured for a cost of about $10,000. This would consist of several panels, making up a total of about 200 meters.

These tests were important because it is believed that the trawl net is an effective countermeasure to the bottom-pressure mine. It is hoped that the Naval Sea Systems Command would be able to fund procurement of some of these trawl nets for use in case of need by the U.S. Navy. See Figure 8.9 for a view of the operational deployment of the trawl sweep net.

Self-Protection Equipment for Mine Warfare Vessels

Replica by Irvin Ltd. of U.K. A radar decoy, launched over the side. It consists of metal-plate radar reflectors encased in inflatable tubes. The radar reflector plates are cut to missile wavelengths so as to be more attractive to enemy missile radars than the ship.

TOAD (Towed Offshore Active Decoy). A jammer, which is activated by a missile's radar. It is towed aft of the ship. Activated by batteries charged by two small windmills and by a solar-energy system. It requires no operator and can be towed by merchant ships.

Matilda. A radar early-warning receiver. Made by the U.K. firm of MEL. Installed in minesweepers to warn them of impending missile attack and of other enemy radar signals.

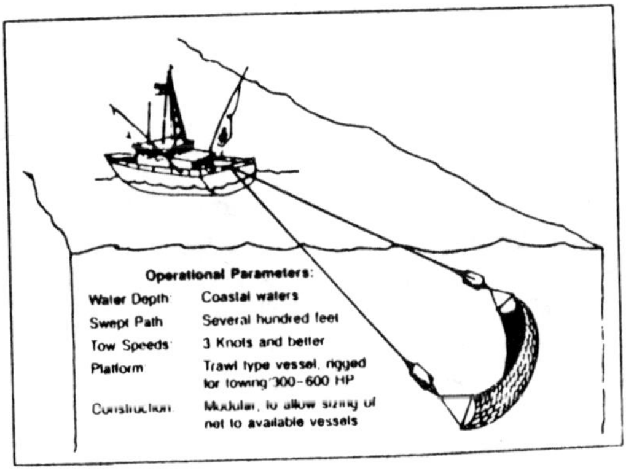

Figure 8.9. Operation of the U.S. Navy trawl sweep. An effective, cheap and simple mine countermeasure.

Acoustic Quieting of Vessels

Part of self-protection equipment involves the systems used to quiet the mine warfare vessels, or for that matter, any vessel that could operate in or near minefields. Ships, within cost constraints, can be made as quiet as possible, both in respect to engines and other machinery, in an effort to reduce the noise they make passing through the water. A great deal of noise is caused by water falling off the hull and particularly the propeller. Efforts are being made by all countries to quiet ships, and special propeller shapes designed to reduce noise are now available. Noise is an extremely complicated matter as the noise created by ships is composed of so many different types of sounds, all of different frequencies, and the mines themselves may be designed to pick out some noise frequencies that are particularly difficult to suppress.

The only self-protection against the pressure mine is to advance at slow speed. Pressure mines are more likely to be actuated in shallow water, where any change of water pressure on the mine's sensors is more pronounced.

One countermeasure against the pressure mine has been the minesweeper special (MSS). These were old merchant ships filled with styrofoam and fitted with very large outboard motors, operated in the 1960s and 1970s. They were manned by minimal crews in shock-mounted pilot houses. Their mission was to clear a channel by detonating the waiting mines.

Today's technology may allow unmanned operation of such vessels. The MSS could transit the channel and quickly clear a safe exit from the port. Its size and bulk are sufficient to provide the waiting mines a full-sized ship pressure signature.

Depending on availability and budgets, each major Navy port could have at least two MSSs. Members of the Selected Reserve could man these vessels. While not a substitute for additional minesweepers, MSSs would be valuable in the early stages of a conflict to assist in the sortie of the fleet from its mine-blocked home ports.

Additional Self-Protection for Mine Warfare Vessels

Some warning equipment and protective weapons, as used on British MCM vessels:

- Barricade chaff-launchers
- ECM warning receivers
- 7.62 mm machine guns
- SCOT Satcomm terminals.

The Soviet Navy apparently places greater emphasis on having guns and missiles onboard for protection against air attack than other navies. The U.S. Navy has very little protection on MCM vessels, in comparison. The British Navy seems to have more interest in self-protection, probably because many of the Royal Navy operating areas would be in the English Channel, North Sea, etc., areas that are quite close to where they could be affected by enemy action against the MCM vessels.

Helicopter Mine Countermeasures

U.S.A. Mk-105 Towed sweeper behind RH-53E. Active acoustic array.
Mfgr: EDO Corp.

ALQ-166 New towed magnetic minesweeper.
Mfgr: EDO Corp.

•AQS-14 Helicopter-towed high-speed minesweeping sonar.
Mfgr: Westinghouse Co.

•AQS-14 is a helicopter-towed side-scan sonar. Normally towed by the RH-53D helicopter, but could be fitted in slightly smaller helicopters. The towed body is 3 m long. Towing cable is 107 m long and is non-magnetic. It is dispensed and retrieved by a hydraulic winch. The sonar is a multi-beam type using electronic beam-forming technology. The control monitor is housed in the right forward portion of the helicopter and is manned by two operators. A tape recorder is used to record the positions of the detected mines.

Helicopters of this U.S.A. type can tow magnetic, acoustic and wire sweeps. They have the advantage of being able to proceed to and from the operational area much faster than can ships, thus saving a great deal of time. Helicopters, however, are not suitable for minehunting and disposal.

The U.S. Navy flew some of these helicopters to the Near East gulf operations area in large C-5 U.S. Air Force aircraft. They were then prepared quickly for sweeping operations. A base either on land or on a ship is still needed to enable the helicopters to be properly supported and maintained.

The Shah of Iran purchased several of the U.S. Navy Sikorsky Sea Stallion helicopters with equipment for minesweeping.

The Japanese Navy has recently purchased several similar but later versions of the Super Sea Stallions for minesweeping operations.

The Soviet Navy uses both the Hi-8 Hip and the Mi-14 Haze-B for minesweeping operations.

Helicopter MCM Operations

The U.S. Navy has increasingly used the helicopter for mine-countermeasures operations. One advantage is its quick reaction capability. It can be placed in a C-5 U.S. Air Force cargo aircraft and flown to an area of operations such as the Persian Gulf in a short time. Of course, either ship or shore support facilities are required from which to operate these helicopters, and this may take some additional time. The favorite U.S. Navy method of mine neutralization or destruction after detection and sweeping is the use of helicopters. The mines, if they are surfaced, can then be destroyed by machine guns.

A towed side-scan sonar such as the Westinghouse AQS-14 can find the moored or bottom-laid mines. Then another helicopter towing an EDO Co. MK-105 or ALQ-166 can destroy or explode the mines with magnetic or acoustic noise. The towed sleds stream a standard magnetic tail astern. Another method is to tow cutters that will sever the moorings of the mines in the water column so that the mines float to the surface, where they can be exploded by gunfire. See Figure 8.10 for an illustration of these sequences. The helicopters (with presently available equipment) are not able to cope with the pressure mine.

A Magnetic Orange Pipe (MOP) is a magnetic mine-countermeasure device that can be towed by helicopters. This is a 10-meter-long iron pipe which is permed with a DC coil to give it an increased magnetic signature. It is filled with styrofoam to keep it afloat and is towed at speeds of 10 to 25 knots. It is normally painted orange.

The helicopters normally carry a Raydist Precision Navigation System or similar system

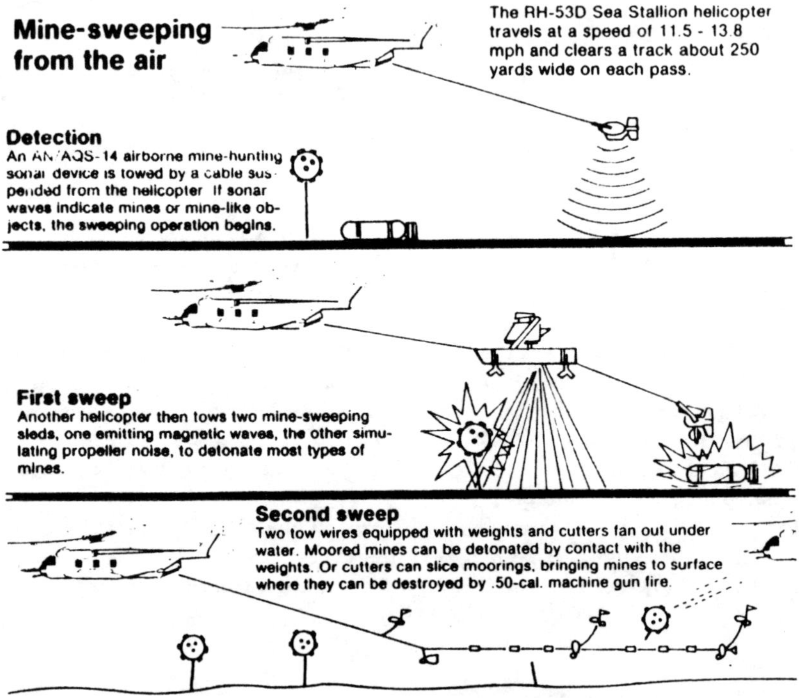

Figure 8.10. Minesweeping from the air with helicopters.

for plotting of position. They can also be controlled by a surface ship radar for positioning.

It is interesting to note that helicopters can be vulnerable to mines. The helicopter rotor noise can actuate acoustic mines of certain frequencies. In some cases, a helicopter can even set off sensitive magnetic mines. A possible enemy counter-countermeasure might be the use of a few "antisweeper" mines scattered in with other mines.

A Unique U.S. Navy Surveillance System—The TSQ-108

MIUW—Mine Inshore Underwater Warfare Unit
IUWG—Inshore Underwater Warfare Group

These organizations are concerned with coastal surveillance. Regular and Reserve U.S. Naval organizations had mobile ISO vans that could be moved to various geographical areas as the missions required to perform surveillance. See Figure 8.11 for an overall view of the operational system. Note that in addition to the radar and communications, sonar-type devices could be deployed to detect underwater swimmers, vehicles and other devices. A number of these installations were procured so that coastal areas could be monitored, perhaps even to detect the enemy laying mines or conducting reconnaissance for intelligence purposes.

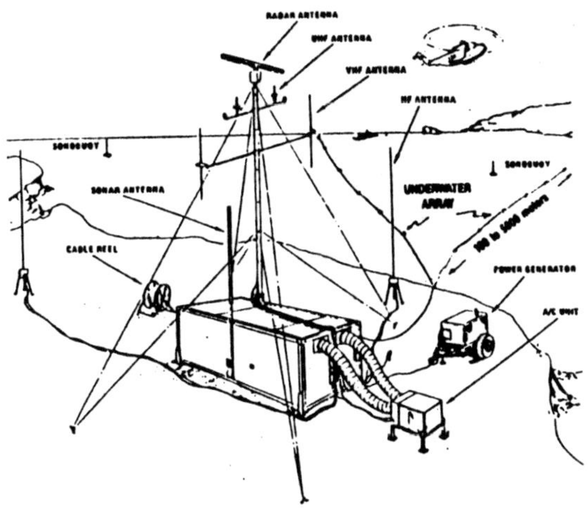

Figure 8.11. U.S. Navy TSQ-108 Radar-Sonar Surveillance Central has mine-countermeasure application.

See Figure 8.12 for a cut-out view of the inside of one of the shelters. Figure 8.13 shows the system communications interface diagram. Figure 8.14 shows an SSQ-58A, which is the standard sonobuoy or sonar listening device that can be deployed and retrieved by an operating crew with small boats. The manufacturer of the SSQ-58A is Sparton Co., which is one of the main manufacturers of sonobuoy and similar devices for the U.S. Navy.

The TSQ-108 mobile installations were designed by the Naval Electronics Command facility in San Diego, California a number of years ago. The Naval Space and Warfare Command took over the Naval Electronics Systems Command mission and facilities.

Use of Divers in Mine Countermeasures

Navy Explosive Ordnance Disposal (EOD) forces are another element of some navies' MCM capability. EOD MCM divers investigate mine-like contacts and can either destroy or recover mines for intelligence exploitation. MCM planners can use this intelligence to utilize limited assets more efficiently, such as targeting minesweepers against a specific type of mine or mine-actuating mechanism.

In the U.S. Navy specially trained and manned EOD mobile units are located on the East and West Coasts in order to provide fleet/force/area commanders with dedicated deployable assets in direct MCM-support roles. These EOD units provide mine readiness certification inspection (MCRI), recovery and scoring services, and harbor channel conditioning requirements. Organizationally, the mobile units consist of some or all of the following:

- Recovery detachments
- EOD response detachments and dedicated MCM detachments

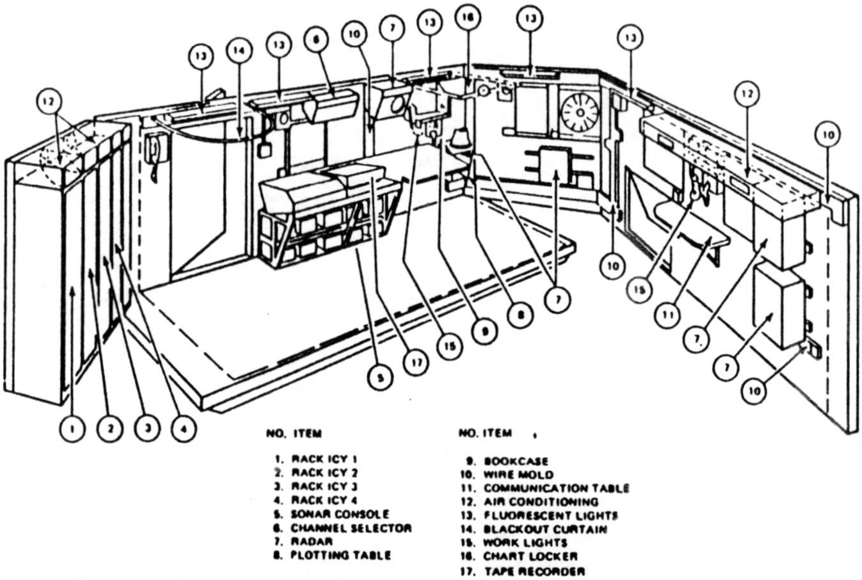

Figure 8.12. TSQ-108 inside view of operation positions and equipment.

The mine scoring and recovery detachments go out during certification inspections of air squadrons and ships to determine if the mines were planted where they were supposed to be and to recover them. They generally deal with inert ordnance. The EOD-dedicated MCM detachments participate in fleet exercises that include mine warfare. They perform integrated tactics with other mine warfare units. The EOD response detachment covers the local area and responds to problems that may arise in areas such as weapon transfers or bomb disposals.

The detachments can use portable sonars and equipment such as the Mk XVI underwater breather apparatus, which can be used to depths of 100 meters. Clearance diving

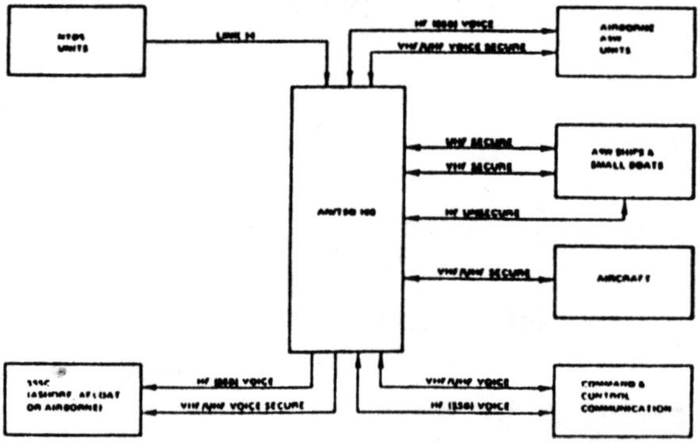

Figure 8.13. TSQ-108 system communications interface diagram.

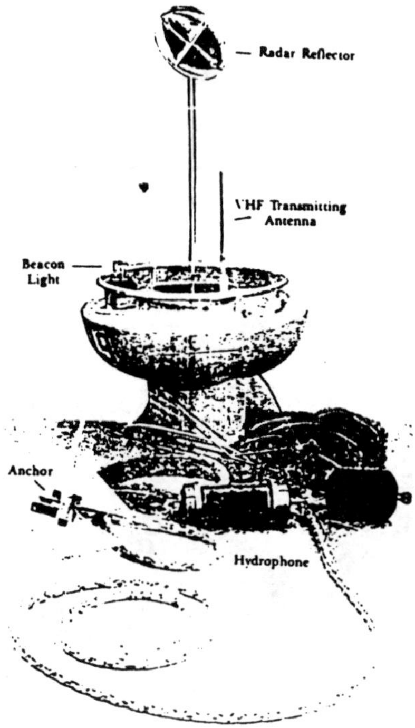

Figure 8.14. SSQ-58A Sonobuoy as used for underwater sound surveillance by the TSQ-108 operations van.

is not an easy task, and considerable risks are undertaken in addition to the normal hazards of diving.

Divers were used for mine clearance of the Suez Canal in Egypt during the 1970s and also in the 1980s, when a unique new type of Soviet-manufactured mine was found by British Royal Navy divers south of the Suez Canal. See Figure 8.15 for a picture of Royal Navy divers using a Swedish "Sea Owl" in a search for mines. Figure 8.16 shows U.S. Navy EOD diving personnel who are participating in MCM exercises. There will probably always be a place for the diver, who may be able to function when operational circumstances forbid the employment of other means.

Diver Minehunting Operations—U.S. Navy

EOD personnel must first detect and locate mines and other underwater ordnance items. Then they can perform the rendering-safe procedures. Mine detection is primarily accomplished by shipboard sonar or towed minehunting sonars and magnetometers. Final classification is made by a diver. EOD diver methods of detection/classification include swimmer hand-held sonars, magnetometers, visual spotting from aircraft and surface craft or underwater vehicles.

Figure 8.15. British Royal Navy divers south of Suez Canal preparing to launch a Swedish civilian "Sea Owl" in search of Soviet mines. U.S. Navy "Seal" teams perform similar functions.

Figure 8.16. Two members from U.S. Navy Explosive Ordnance Disposal Team work from an inflatable raft while participating in mine-countermeasure exercises.

For locating underwater ordnance and mine-like objects, EOD relies on the PQS-1 hand-held active-passive sonar, Mk-9 all-metals-locator device, Mk-10 flux-gate magnetometer, Cubic-Autotape Navigation System DM-40A, Klein Hydroscan sonar Model 400 or later system, and a Magnetometer Sensor System. Mk-16, Mk-71 and Mk-72 acoustic receivers are also used when items which have an active pinger installed are being searched for. The detection range of the PQS-1 sonar is about 110 meters; the Mk-9 detects metal at 1 meter, somewhat less if the object is buried; and the Mk-10 detects metal objects at 3 meters, with a capability of detecting items that are buried at a depth of about 2 meters.

EOD divers can aid fleet commanders in identifying enemy mine characteristics. The best information is obtained when the EOD team renders the mine safe. It can be sent to a Navy Explosive Ordnance Disposal Facility where it can be thoroughly investigated, and for an exploitation of the mine mechanism for coming up with a countermeasure.

EOD divers can neutralize detected mines directly or can assist in the vectored-boat technique of neutralizing the mine. The minesweeper sonar with mine-countermeasures ROVs such as the MNS can be utilized if this is more appropriate or expeditious in a particular operational situation.

For mine retrieval—when it is desired that the mine be exploited for intelligence purposes—EOD personnel operate a remote lifting balloon that is capable of lifting 400 kg of submerged weight from a depth of about 45 meters. If a larger weight is encountered, or a greater depth, two or more lifting balloons may be used simultaneously.

EOD Equipment

Mk-26 Mod 0 Tool Set—designed for underwater ordnance and includes mine and torpedo tools made of non-magnetic alloys. The tools are lighter, stronger, more durable and magnetically safer than older tools. The tool set includes:

- Hand tools
- Adapters
- Special mine tools
- Probe
- Cavity charge containers

Mk-27 Mod 0 Tool Set is designed to support Mk-26 Mod 0 tool set with general non-magnetic hand tools such as crowbars, wire brushers, wrenches, picks, screwdrivers, etc.

Mk-9 Mod 0 Ordnance Locator is a portable, diver-carried electromagnetic-type locator that can detect all metal objects, ferrous and non-ferrous, at a range of about 1 meter. It is capable of operating to depths of 30 meters.

Mk-10 Ordnance Locator is a transistorized gradiometer magnetic-anomaly detector used to locate hidden or buried ferrous objects. It is completely self-contained, portable and designed for hand-held operation in all orientations and background fields. It can be used in fresh or salt water as well as on land. It weighs 5 kg in air and $\frac{1}{4}$ kg in water.

Later locators use Cesium Vapor or Xonics devices. They have an increase in sensitivity and an increased range of detection of mines.

Interval Timers are used by EOD personnel to permit them to leave an area prior to a scheduled detonation either underwater or on land. These timers are waterproof/pressureproof devices, adjustable in intervals of time.

Incendiary Torches are used by EOD personnel as a cutting devices for use underwater or on land to penetrate a mine case and for burning out the main charge contained in the case. The remote opening incendiary torch is capable of penetrating 2 cm of mild steel in two seconds. It is usable to 1000 meters and will sustain combustion of the main charge at this depth.

Underwater Shaped Charges are used by EOD personnel to render explosive ordnance safe. These are linear and conical preloaded charges for mineshell cutting. These have a built-in, water-tight, stand-off feature for protection of the diver, and also have good reliability of the charges.

Mk-18 Marine Mammals Minehunting

Dolphins are being used by the U.S. Navy in searching for mines. Research has been going on for a number of years to verify that the dolphin uses a clicking type of self-generated noise as a sonar to locate objects. Navy facilities in Hawaii have trained dolphins in mine-location procedures. It takes up to two years to train dolphins in these procedures, with an estimated cost of $200,000 for each one trained.

The mammal caretakers, or Navy Special Warfare personnel, move the dolphins to an operating area in special transporters. They can be transported in USAF cargo aircraft such as the C-141 or C-5. The dolphins must be sprayed with water periodically while being transported.

For a typical operation such as neutralizing bottom mines, an Mk-81 "Yellow Banana" device is used. Marking of mine-like objects is done with plastic probes and a sonar reflector. The dolphins are trained to give indications such as:

- When dolphin touches one ball on right—means no mine.
- When dolphin touches one ball on left—means mine located.
- The dolphin then places neutralizing charge or device near the mine.
- The sonar reflector marks the mine location. Forty lbs of C-4 explosive can be used to rupture the mine. The Mk-48 timing clock is used to time the explosion.

Rubber boats that hold two operators can be mounted on a frame for air transport by helicopters to operations areas.

Research and development has been conducted by some Navy laboratories in using the Baluga whale to recover torpedoes and missiles after firing. The whales attach a snap-ring device to the object they are sent to recover.

Soviet Mine Operations

The importance of the mine in the Soviet naval arsenal today is best indicated by the fact that all major fleet units, including new cruisers, are configured to lay mines. Soviet submarines are capable of laying mines and, together with aircraft, can provide the most usable delivery means to plant mines offensively. Mine warfare training is a standard course of instruction for all Soviet naval officers and most Soviet enlisted personnel.

During the Korean War at Wonsan Harbor, the Soviets did not employ their main Navy to conduct a mining campaign. With Soviet-produced mines, a few junks, wooden barges, and sampans, they were able to accomplish the mission. About 3000 mines, both contact and influence types, were laid in Wonsan during a three-week period. Some of these Soviet mines could even be detonated by the magnetism of machinery and hull-fittings of wooden minesweepers! This resulted in the loss of some minesweepers.

According to Soviet philosophy, there are three types of mine barrier that can be employed. The first and possibly most widely used type of barrier is the general moored or ground minefield used to cover a wide coastal area with mines, usually of one type, but laid with varying intervals between lanes and mines and with areas of the field extending for various distances with mines laid at varying depths. In mid and deep water in open ocean areas, the barrier made up of one or more lines of mines of different types at set distances is more common. Finally there is the barrier used to block straits and choke points and the approaches to bases and ports. This barrier consists of the mine cluster, which is a series of obstacles about 1 km in length, with a small number of mines laid in a line or a field with different intervals and distances between individual mines.

Minesweepers were initially used in Port Arthur during the Russo-Japanese War of 1904-1905. At the time, they were actually auxiliary ships adapted for minesweeping. The Russians believed that special ships would be required to fight mines successfully. Therefore, a new type of ship was built in 1909 in St. Petersburg. They were called "Minrep" (mooring line) and "Vzryv" (explosion). Minesweeping brigades took part in World War I and World War II operations.

Soviet minesweeping personnel took part in demining the Suez Canal and ports in Bangladesh. They operated under extremely difficult conditions, far from home bases and support. A great number of seamen were decorated with government awards. See Figure 8.17 for a view of a typical Soviet minesweeper. Note the sweeping device suspended from the crane on the stern of the vessel. Figure 8.18 is a picture of Senior Lieutenant S. Porkofyev, the commander of the minesweeper of the Soviet Northern Fleet. The Soviets give awards for the best minesweeping teams.

Modern Soviet minesweepers (ocean, coastal and harbor alike) are provided with echo-ranging sonars and television equipment to search for and classify the detected mines,

Figure 8.17. A typical Soviet Navy minesweeper. Note the sweeping device suspended from the crane on the stern.

Figure 8.18. Soviet Senior Lieutenant S. Porkofyev, Commander of a Soviet Navy minesweeper.

sweeps for different purposes, demolition charges for mine blasting, and remote-control apparatus for mine detection and clearance.

Minesweeping Operations

To locate a minefield, determine its boundaries and establish the degree of navigation safety needed, an exploratory sweep and then a test sweep are performed with the aid of contact and influence sweeps. Top priority is given to inspecting the areas most essential for navigation. The mines detected are recorded and may be neutralized later on.

Contact sweeps are used in fighting moored, floating and drifting contact and influence mines, and anti-minesweeping explosive floats. The latter include explosive obstructors breaking the sweep or paravane with special cartridges, and cutter-type floats severing the sweeping section with special cutters. The main part of a contact sweep is a steel rope (sweeping section) kept at a predetermined depth with the aid of buoys and special devices. As regards their effects on mines, such sweeps are subdivided into:

- Cutting—breaking the mooring line with the aid of demolition charges or cutters.
- Towing—gripping the mine mooring line and taking it together with the anchor to shallow water for subsequent deactivation.
- Marking Sweeps—designating mines with stakes or buoys.
- Net Sweeps—nets secured on a net frame. (Used for picking up floating and broken-adrift mines, and ground mines planted on a hard bottom.)

Influence Sweeps

Meant for disposal of ground and moored influence mines by acting on their timing devices with simulated physical fields such as:

- Acoustic
- Hydroacoustic
- Electric
- Electromagnetic

- Magnetic
- Hydrodynamic
- Thermal
- Gravitational

As to the effects that they produce on mines, influence sweeps may be of electromagnetic, acoustic and combined types. In their turn, electromagnetic sweeps are subdivided into loop, solenoid and open-circuit sweeps, depending on the method used to generate the magnetic field. The sweeping section of a loop sweep is a ring electric circuit, and that of a solenoid sweep, a 15-25-meter-long iron core with a winding through which an electric current is passed. The sweeping section of an open-circuit sweep comprises two cables terminating in electrodes, with electric current between them floating through sea water. The power consumed by the sweeps ranges from 50 to 600 kW.

Depending on the frequency range of the simulated field, acoustic sweeps may be of a narrow- or wide-band type. An acoustic sweep is actually a towed buoy housing a percussion mechanism to generate an acoustic field or noise in an aqueous medium. There are also acoustic sweeps of an explosive type, consisting of a water-submerged tube through which small-weight explosive charges are dropped. There are also hydrodynamic sweeps imitating changes in the water pressure brought about by a moving ship. Combined sweeps include jointly towed electromagnetic and acoustic sweeps, and also sweeping boats and mine bumpers.

A sweep is generally made up of an outboard part and shipboard equipment. The outboard portion consists of a towline, a depressor, a sweeping section (search elements), otters, marking buoys and paravane floats.

The shipboard equipment consists of minesweeping winches, anchor davits (cranes), electric-current generators for powering the sweep, converters, monitoring and remote-control devices, etc.

Sweeps are used in combination with mine detectors, which are actually towed or remotely controlled self-propelled systems intended for mine location, marking and disposal. Depending on the physical principle underlying the search method, mine detectors are divided into hydroacoustic, optical, magnetic, electromagnetic, electric and laser types. In hydroacoustic detectors, the energy radiated by its source is reflected from the underwater object, to be recorded by a special receiver. Optical detectors locate underwater objects with the aid of TV and photo devices. Magnetic, electromagnetic and electric mine detectors employ the principle of registering the changes in the respective physical fields with mines getting in the way of the latter. A laser beam is used in laser detectors to identify the submerged object.

Mines are swept up by groups or divisions of minesweeping ships. When mine characteristics are unknown, explosives or mine bumpers are generally used.

Figure 8.19 is a view of a Soviet double-cutting sweep mechanism.

Other Soviet Mine Operations

The Soviet Navy has used Mi-8 "HIP" helicopters deployed from the cruiser *Leningrad* in the 1970s operations in the Strait of Gubal, in the Gulf of Suez. The helicopters were used in mine-countermeasure operations, after the 1973 Arab-Israeli war.

In 1984, the Soviet helicopter carrier *Leningrad* in company with a "Kashin" class destroyer conducted mine-clearing operations in the Southern Red Sea. On her flight deck were two Mi-14 Haze-B mine-countermeasures helicopters. These two helicopters did not form part of the ship's normal air complement. They had to be kept on the

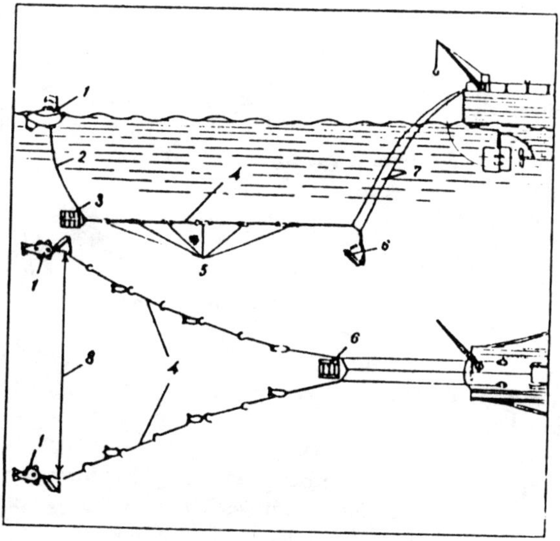

Figure 8.19. Soviet Navy double-cutting sweep.

1. Buoy
2. Buoy cable pendant
3. Otter
4. Sweeping Section
5. Cutters
6. Depressor
7. Towline
8. Grip width

flight deck under weatherproof tarpaulins because they were too large for the ship's elevators.

On most of the Soviet minesweepers in the Gulf of Suez during these operations, the ships' crews had received previous minesweeping task training during exercises and inspections. But in a real environment—when the mines were real and live—the experience level was very shallow on the Soviet ships. However, the officers and warrant officers who had wartime training quickly trained the crewmen. Conditions such as the terrible heat of the latitudes of the detachment's course (among the hottest on the planet) and lack of fresh water and raging sandstorms were continuing problems. Inaccurate maps caused difficulties with the coral reefs and the complex relief of the bottom in the Gulf of Suez. There were very poor records available from the Egyptian Navy about where the Israelis had laid the mines. On one occasion, a bottom mine blew up off the side of a Soviet minesweeper. The ship was damaged and water began to fill both engine compartments. Fortunately, the damage-control party brought the situation under control. On another occasion, while the ship was in a minefield, an Israeli gunboat with uncovered large-caliber machine guns circled her, and then simulated a clash, attempting to move the Soviet minesweeper off its combat tack and into coral shallows. The Soviet commander, however held his course and the gunboat took off at high speed.

CHAPTER 9

Outlook for the Future

GENERAL

The late Rear Admiral Charles W. Cummings, upon taking charge of the U.S. Navy Mine Force in the summer of 1974, expressed surprise and dismay at the vast amount of technology with possible mine-countermeasures applications that was either unnoticed or ignored by mine warfare "traditionalists." For example, lasers might be of use in both detecting and destroying mines. Infrared radiation may be useful in detecting mines, and electromagnetic radiation could possibly "zap" circuitry and render mines inoperative. As Admiral Cummings pointed out, "The technology exists."

The United States of America is today more vulnerable to a mining campaign than at any time in the nation's history. With many of our resource needs being transported by sea and with our fleet concentrated at a relatively small number of bases, even a relatively crude mining operation could alter the balance of power. The mines could even be laid covertly by an enemy and then be commanded to turn on by remote control at a later date. What do not yet exist are the priorities, organization, and funding necessary to achieve a truly reliable and efficient detection and neutralization system.

Note a recent comment from an Industry "EIA" conference: "Navy mine warfare is receiving considerably more attention, a trend that it is hoped will continue. However, no evidence is found as yet that mine warfare is being elevated in real priority."

In 1982, it was said, "Mines may be weapons that wait, but mine warfare will wait no more." Former CNO Admiral Thomas B. Hayward put high emphasis on mine warfare. Later CNO Admiral James D. Watkins continued the momentum by assigning a high priority to mine warfare. New mines were coming into the Fleet, new mine warfare ships and helicopters were programmed, and new mine-countermeasures systems were in the budget. Figure 9.1 shows the new MH-53E MCM helicopter "Super Sea Stallion" that is now entering the fleet. Figure 9.2 shows that the U.S. Navy even experimented with the use of surface combatants to lay mines. The *U.S.S. Meyerkord*, FF-1508, was configured with a special rack on its stern to launch mines. This was a good step forward and a fine improvisation, but there seemed to be no follow-up to apply it to fleet use. The sad attempt to build a surface effect mine-countermeasures ship called the MSH—minesweeper/hunter "Cardinal" class set the U.S. Navy program back. Italian GRP

Figure 9.1. U.S. Navy new MH-53E Minesweeping "Super Stallion" helicopter.

Figure 9.2. *U.S.S. Meyerkord*, FF-1058 configured with a prototype mine-launching system on its stern—off the side of the helicopter landing pad.

technology and hull design are now being utilized in a renewed attempt to produce the MHC-51 coastal minehunter similar to the Italian Navy "Lerici" class.

In 1981, Admiral Charles F. Horne III, Commander of the Mine Warfare Command in Charleston, South Carolina, said:

> "In all of recent history up to now, approximately one and one-third million mines have been employed in warfare, and since the Civil War, naval mines have been responsible for the sinking or damage of more ships than any other single weapon system."

Given this statement, it is disheartening to know that the U.S. still lags behind the Soviet Union when it comes to offensive and defensive mine warfare capability. Former Secretary of Defense Casper Weinberger, in his FY83 Defense Posture report to Congress, warned that the U.S. mine warfare inventory would be only "marginally effective against the Soviet threat."

There is a turn-around on the emphasis needed to attract young talent to the mine warfare force. The mine warfare job should not be an unglamorous and low-tech mission, nor one to be escaped at first opportunity. A career must be enhanced for one to stay in the mine warfare career field. As Capt. Robert H. Smith, U.S.N. Retired, once said: "It is a familiar and yet shallow truism that the U.S. Navy praises mines in concept even as it neglects them in practice."

As proof of the U.S. Navy's low interest in mine warfare, critics point to the fact that virtually all of the current minesweeping ships are assigned to the reserves and the new ships are scheduled to shift to the reserves after about a year in the regular fleet. History shows that the Navy, and most members of Congress, become interested in naval mines and the countermeasures only when the issue is brought to their attention by a conflict.

The "Super Coop" idea evolved from the Navy's Craft of Opportunity Program (COOP), which called for the Naval Reserve to receive 22 YP-683 ships—outfitted with limited minesweeping capabilities—for training and mapping harbors. The Naval Reserve Association, along with a number of congressmen, liked the idea, but proposed that the Naval Reserve be given the YP-683 vessels equipped with both mine-hunting and mine-neutralization equipment. The U.S. Navy has however resisted the Super Coop concept. The special YP-683 ships would cost much less than MCM and MHC ships, but would be just as effective as the more expensive ships in harbor and shallow-water environments. Apparently Congress and the U.S. Navy have battled over this issue for the past several years. Is this another case of "foot dragging"?

With the recent constraints on spending due to the effects of the Gramm-Rudman environment, the defense spending will impact heavily on U.S. Navy spending. This in turn means that the priority will be placed on submarine, aviation, aircraft carrier, and missile systems, and not necessarily on mine warfare. So the future is not rosy for the mine business. It will be up to a few key admirals in critical positions; such as NAVSEA and OPNAV to keep pressing for funds for the mine warfare people and hardware.

The Navy's maritime strategy stresses the critical need for highly capable mine warfare assets to conduct mine-countermeasures (MCM) operations and offensive and defensive mining of important sea areas. Unfortunately, the U.S. Navy lacks adequate forces to protect even our own coastal waters and harbors, let alone project MCM platforms into regions where operations will be conducted. Therefore, perhaps we are relying on the mine warfare assets of our allies to clear ports, harbors, and critical sea lanes in the event of war. Even our allies may let their mine warfare assets dwindle to a point where they can help the U.S. very little. Recent events have shown that it is difficult to muster allied support for collective MCM operations in areas far removed from traditional areas of defense cooperation—particularly in peacetime crisis short of war. Even U.S. friends in the Persian Gulf operation region were reluctant at first to show visible support for the U.S. Navy, which should have been in their own interest.

The real picture of mine warfare in the U.S. Navy is that it has a low priority and is the first to be cut when budget reductions must be made. This level of interest can be traced back to Alfred Thayer Mahan, the founder of doctrine of the modern U.S. Navy. He disliked the mines, calling them an unnecessary burden on a Navy intent on achieving command of the sea. Mahan believed that mines constituted a mode of warfare that those who commanded the seas did not want, and which if successful would deprive them of it. The British admirals of the time believed the same, and perhaps this influenced him.

Very few of the active-duty naval officers who were in mine warfare positions considered themselves to be career mine warfare officers. A long association with mine warfare was not career-enhancing. Some officers who wanted a command early on sought to get into mine warfare, but after the command they moved on to something else. Often the mine warfare person was primarily a helo, S-3 or P-3 pilot, an ASW surface type or submariner, in their own eyes, and they soon looked for greener career pastures.

The bulk of the mine warfare personnel are naval reservists and are dedicated professionals who try to do a good job. During the recent Gulf crisis of mines, a number of mine vessels were needed in a hurry. The reservists could have been called up for this duty. Unfortunately, both COMSURFPAC and COMSURFLANT looked to other fleet ships to gather crews to send to the crisis area. There should have been enough active naval reservists available to fill the billets.

One should contrast this method of operating to those of the mine warfare forces of our allies and of the Soviet Navy. For example, the British and Germans consider that mine warfare is a career profession.

Often Europeans think that the U.S. takes a "childish, amateurish" view of mine warfare. They think that the U.S. is not aware of the severe threat that there is to U.S. shipping if the Soviets were to deploy their mines. French, Belgian and Netherlands mine warfare officers are of the same opinion.

As a result of a decision made during the Admiral Zumwalt era, airborne mine countermeasures (AMCM) received priority. Funding was provided for more MH-53E helicopters to pull sleds through the water. Most helicopter pilots would probably be in a more career-enhancing billet such an ASW. All they seem to do is drag metal through the water and fight boredom.

By contrast, the Soviet Navy has consistently maintained strong mine warfare forces. Soviet doctrine is that mines, like missiles, have become so effective that today they comprise an essential part of modern warfare in all maritime operations. They regularly practice mining and MCM (mine-countermeasure) operations. The Soviets have the world's largest stockpile of naval mines; approximately 400,000 mines of all types, including many with highly sophisticated capabilities. They believe in mass deployment of mines at the outset of a war. Soviet defensive mines would probably be deployed to protect their ballistic missile submarine bases. If the U.S. sends its nuclear-powered attack submarines (SSNs) and several carrier battle groups into these areas as is contemplated with the forward maritime strategy of the U.S. Navy, there might be a serious threat to them from these Soviet defensive minefields. U.S. ports would probably be mined, as well as some choke points. This would pose a serious problem for logistics planners who plan to send strategic sealift vessels over to Europe early, within ten days of the start of hostilities, with critical material. Without this surge of shipping, NATO could not sustain military operations for more than two or three weeks under most scenarios.

Oversight of mine warfare programs is not centralized in the U.S. Navy. There is no focus other than the Secretary of the Navy, Chief of Naval operations or Vice Chief of Naval Operations. During the Korean War, 1950 to 1955, there was a big minesweeper building boom. In 1977, Vice Admiral James Doyle, the Deputy Chief of Naval Operations for Surface Warfare, attempted to get funding for a deep-ocean MCM ship program into the 1979 Navy budget. This commendable effort turned into the "Avenger" new ship program of the early 1980s.

Within the Navy itself, there is no single focal point for mine warfare. Both Deputy Chiefs of Naval Operations for Air and Surface Warfare oversee their parts of the program, and program execution and research is done at separate offices in Naval Air and Naval Sea Commands, Navy laboratories, centers and activities. Even the day-to-day administrative command of the Navy's active and reserve mine warfare forces is divided. Surface assets are under the Commander, Mine Warfare Command in Charleston, South Carolina, and responsibilities for the MCM helicopters is split between the Commander Naval Air Force, Atlantic Fleet at Norfolk and the Pacific Fleet at North Island, California.

Mine warfare has enjoyed a boost in the early and mid-1980s, benefiting from the increased defense spending of the Reagan administration. With the ensuing budget cuts, there seems to be little hope that this emphasis can be sustained in the next decade. A lot needs to be done to protect the successes achieved so far.

Recent problems with the "Avenger" new-ship construction have cast a shadow on our attempts to build a new mine surface ship force. The ship designers did not allow for proper integration of ship's engines to the drive mechanism, so much valuable construction time was lost to correct the problems. Even the catamaran-type hull mine vessel being built for the Navy was cancelled because of "technical" problems with its fiberglass hull during tests. Allegedly, the near-explosion tests conducted on the hull were too severe on the ship.

As a stopgap measure, the U.S. Navy has had to resort to using the "Lerici" Italian fiberglass hull conventional design for a mine vessel. Some consideration could have been given to a non-magnetic steel design such as is used by the German Navy for new minesweepers. Apparently this was not possible because it was believed the U.S. shipbuilders/steel yards did not or could not master the skills for welding this special steel—an unfortunate circumstance if it is true, since it does not speak well for our technology in the U.S.A.

The Soviets for many years used mines for purely defensive purposes. They recently have expanded their plans for use of mines in a worldwide offensive way. The United States has realized that certain maritime gateways of the world are more vulnerable to mines than they are to sabotage or terrorism. Typical of these are the sea route around South Africa, the Panama and Suez canals, the Caribbean passages, the Strait of Malacca and the Persian Gulf and its approaches.

The Soviet Union has a strategic geographical disadvantage. They have no direct outlet to the most important ocean, the North Atlantic. From the port of Murmansk on the Arctic Ocean, they must run the NATO gauntlet off the tip of Norway and past the flanks of Iceland. The Baltic Sea and Black Sea outlets from Leningrad-Talinn and Odessa-Sevastapol are controlled by various of the NATO countries. The Pacific Ocean outlets for Vladivostok-Nakhodka through the Sea of Japan are limited both by geography and winter weather off Japan, with Japan being a U.S. friend. Allied strategic plans should therefore call for denying these routes by use of mines, and that means exploiting the Soviet vulnerability. Offensive minefields complement other sea and air control forces.

The U.S. Navy then, should have shallow-water, medium-depth and deep-sea capabilities in mine planting. The readily accomplished airborne mining of Haiphong Harbor in North Vietnam encouraged this type of mine warfare development. The

U.S. Navy is also affected by the U.S.S.R. and its Warsaw Pact allies. Soviet bottom-influence mines are a threat to mine-countermeasures ships of the NATO allies, and the ability to mine deep-water shipping lanes and narrow passages poses another problem to be solved.

Some Considerations for the Future—Concepts

In any conflict, MCM ships will be considered prime targets for attack by surface ships and aircraft. It is essential, therefore that they be provided with some form of electronic support measures (ESM) receiver, coupled to a chaff decoy dispenser. It is significant that the British "Hunt" class MCMVs going out to the Gulf in 1987 were hurriedly fitted with MEL Ltd. Matilda radar threat warning receivers. The British had learned their lessons well from the war with Argentina in the Falklands/Malvinas.

In any conflict, MCMVs may be expected to and should be capable of undertaking long ocean passages. That is not so today, particularly in the case of the older U.S. Navy MSOs. Habitability is hard to achieve in very small ships such as MCMVs. The support organization for sustained MCM operations must allow for a frequent exchange of key personnel and adequate rest periods for everyone.

It seems inevitable that MCM will continue to be led by developments in mine technology, but the gap can be kept narrow. Two important considerations in the near future will be deeper-laid mines and the "intelligent" self-propelled mine, both of which will pose serious problems. Another refinement could be the counter-attacking mine, possibly programmed to react to a minehunter's sonar. Another possibility is tethering moored mines to react to a mine-disposal vehicle; a chain reaction of explosions might easily destroy the minehunter, as well.

On the positive side, more efficient sonars and command and control systems will become available. Automatic tracking will reduce the workload on operators. MCM air-cushion vessels should be very useful in MCM work if designed properly.

Mine-Avoidance Submarine-Tethered ROV

There is a need for a remote operating vehicle (ROV) that will operate from and forward of a submarine to ensure an essentially safe path through a minefield.

The ROV should consist of a sonar transducer and hydrophone array operating at multiple frequencies in order to provide long-range search, medium-range detection, and short-range classification capabilities. The ROV would be tethered forward of the submarine at a predetermined distance to detect and classify mines and relay the information by cable to a display onboard the submarine, thereby allowing time to maneuver. The coverage of the sonar set would be 180 degrees forward of the ship, with the capability of side-scan for 270-degree coverage. The system could operate narrow-band for high resolution, and would provide a display of range, height-above-bottom, and three dimensional imaging. The ROV could be stored on the submarine when not in use. One concept would be for the vehicle to be torpedo-sized, and therefore it could operate from an open torpedo tube. A fiber-optic tether cable could be used to carry sensor and control signals. The threat to our submarines from mines will continue to grow as smarter mines are developed. It is time now to develop the smart submarine anti-mine system. The best in anechoic coatings, towed array sonars or digital fire-control systems are helpless against a well-placed rocket-propelled rising mine!

MADOM Program

A multi-sensor mine detector, which is to be towed behind the mine-sweepers. This is an FY 1990 new start with a planned procurement of 30 or 40 systems. In the future, a version of this system could be used by the U.S. Marines in sweeping shallow waters prior to amphibious operations.

AMDAS

A laser sensor system that is to be used on airborne platforms to sweep for shallow-water mines. This system does not have to be towed through the water, which will allow for much faster sweep rates.

Organic Minesweeping for High-Value Navy Ships

Thought should be given to equipping high-value ships with an organic onboard minesweeping capability. Each ship could have its own UUV (Unmanned Underwater Vehicle) minesweeper that would test the waters and map out a pass through the minefields.

BURMMS

Buried-Mine Minehunting System. Development and field testing of such a system has begun. Naval Sea Systems Command and Naval Coastal Systems Center have funded and are sponsoring the work. Westinghouse Co. of Baltimore, Maryland is involved. The instrument prototype would be suited to hunting electromagnetically underwater for mines buried in the sea bottom. Physical Dynamics Inc. of San Diego, California is also involved.

Nuclear Mines

Both superpowers, the U.S.A. and U.S.S.R., are known to have nuclear sea mines. It is hoped that neither would decide to use them.

Advanced High-Technology Mines

The ability of mine manufacturers to introduce new types of mines should never be underestimated. Mines can be made invisible to sonar. For instance, recent developments have included mines with a covering that absorbs ultra-sonic rays, mines of special shapes, and mines incorporating a vibrating mechanism to bury them in the sea bed. Of course, sonar manufacturers are aware of these developments and are trying to produce equipment capable of overcoming them.

Advanced Robots—SPUR

The European-based Scicon Co. has conceptually developed SPUR (Scicon's Patrolling Underwater Robot). The full-scale vehicle would be about 10 meters long with a 1.8-meter cross section. Normal propulsion would be by an oxygen/hydrogen fuel cell, and higher attack speeds of up to 50 knots may be obtained from either a closed-cycle engine or batteries. At cruising speed, the vehicle would have an action radius of

1850 km, an endurance of two months and, for certain applications, be capable of diving to 6000 meters. SPUR would be under shore control and artificial intelligence would be used to assist functions which SPUR would need to perform autonomously, especially in the fields of tactical decision making, such as:

- Route planning
- Target classification
- Attack maneuvers
- Communications routines

Roles for the SPUR could be:

- Mine countermeasures (Using manipulators)
- Vessel destruction (Via torpedoes)
- Wrap-around wire system (Directed against ship's propellers)
- Intelligent mobile mine

Note: Possibly "Wolf Packs" of SPURS deployed in a patrol line to create barriers up to 370 km long could be practical.

Future Platforms for Mine Warfare

A growing threat to NATO is the Warsaw Pact's ability to use commercial vessels for covert minelaying before war has broken out. Ro-Ro type ships are ideal for this kind of operation. Indeed, Libya may have used such a vessel to covertly lay mines by rolling them off the stern ramp in the Near East in the 1980s.

The Norwegian Navy has allocated a number of Ro-Ro ships to defensive minelaying in time of crisis. Other nations may be doing likewise.

The C-130 Hercules four-engine cargo aircraft which have been sold to many nations by Lockheed Corp. of the U.S.A. is an ideal aircraft for minelaying. The aft cargo door is used to launch the mines.

The airship or "blimps" as they are sometimes called, could be used as a platform to either plant mines or to tow mine-sweeping devices that could detonate mines without endangering the airship. Flying at 30 meters above the water, the airship would have long endurance time on station and many conveniences for the crew. The airship is unlike the helicopter in that it does not use large amounts of fuel and power to remain stationary or when sweeping mines.

Satellites will play an increasing role in the mine warfare. This will be primarily in the navigational field. More exact navigation will be possible by utilizing the high accuracy of the Sat-Nav system.

Future ROVs will range from today's submersible "mini-subs" tied to a mother ship, or operating autonomously, to ocean bottom tethered tractors.

The U.S. Navy is continuing to build wooden mine warfare ships. The main effort is with the MHS "Avenger" class vessels being constructed by both Peterson Builders Inc. and Marinette Marine in Wisconsin, U.S.A. However, considerable effort has been put into improving the construction facilities and speeding the fabrication process. Figure 9.3 shows one of the MSH class wooden vessels being constructed in the Marinette Marine Co. facility. By the use of mezzanines, workers can reach any point on the ship under construction easily. This facility has been constructed with the thought in mind of being able to construct wooden, steel or fiberglass hull vessels. The building in which the ships are built is environmentally controlled. This is an advantage because the winters

Figure 9.3. Marinette Marine Co. new facility for producing wooden, steel or fiberglass mine vessels. Mezzanines enable workers to reach any point on the ship under construction. Shown is one of the U.S. Navy MSH "Avenger" class vessels under fabrication.

in Wisconsin are terribly cold. A GRP (Glass Reinforced Plastic) sheathing is used over the wooden hull planks to protect the hull.

Dependable, cost-effective, and rapidly produced vessels hold the key for the U.S. and its allies in the next decade. One must keep in mind that we may not have the time to build a quantity of such vessels in the next conflict!

ALQ-166 New EDO Co. Minesweeper

The EDO Co. of College Point, New York is now providing the U.S. Navy with the next generation of airborne, helicopter-towed magnetic minesweeping equipment. See Figure 9.4 for a view of the sweeper under tow. Hydrodynamic, electronic, power, deployment and handling tests have been successfully accomplished. This new high performance system provides a significant increase in capability over the older EDO Mk-105 system that was used in the Haiphong Harbor and the Suez Canal.

Fiber Optics

More use will be made of fiber optics in cables for tethered ROVs. Also sensors can be made using fiber optic techniques. Optical fiber is ideal for use because of its immunity to interferences, either incidental or man-made, and large bandwidth capability with small physical size and weight.

Figure 9.4. The new U.S. Navy ALQ-166 helicopter-towed magnetic minesweeper.

Mine Wareface Enhancements by Modifications

When cost is a problem, and time of the essence, it is often possible to change existing systems in a way that greatly improves performance at a reasonable cost and in a timely manner. This is sometimes the case with aircraft or tanks where a new engine can be installed, a new gun or missile mounted on an already existing carriage assembly, etc. France has done this recently by modifying an already existing DUBM-41 towed sonar to study the feasibility of detecting buried mines. (See figure 9.5.) Thomson-CSF underwater systems branch Thomson-Sintra is under contract to the French Admiralty Group d'Etudes Sous-Marines de L'Atlantique (GESMA) for this work.

Likewise, more than just lip-service should be paid to the Mine Warfare COOP Programs – Crafts of Opportunity that can be outfitted to assist in the mine warfare business when the need arises. Some of the European and even Third World navies are interested in using the concept, and have done some developmental work. Properly thought out and implemented, the COOP mine warfare program should contribute meaningfully to the U.S. Navy mine warfare capability.

Future Soviet Advancements in Mine Warfare

What the Soviets write about is indeed borne out by the way that they equip their ships and aircraft. Old cruisers with large numbers of mine rails are retained in service long past the date they would be retired from a NATO navy. New small and medium

Figure 9.5. Mine detection sonar built around modified DUBM-41 towed sonar of the French Navy.

Soviet combatants are entering service with installed mine rails. This capability is totally lacking from even the oldest active U.S. Navy combatants. All Soviet submarines, both diesel and nuclear-powered, are said by western analysts to have both a minelaying capability and mission. Many Soviet Naval Aviation "Backfire" bombers have been photographed with what appear to be mine racks under their forward fuselages.

To assume that Soviet capabilities in mine warfare will remain static would be a definite mistake. Captain First Rank V. Yankovskiy, a Soviet author writing in their Zarubezhnoye Voyennoye Obozreniye (Foreign Military Observer) paper spoke of the development of a "self-transporting mine" that could be used to mine harbor approaches, straits, narrows and shallow channels. This is probably a stand-off weapon and follows the Soviet early development of cruise missiles. It would be very effective in the anti-SLOC (Sea Lines of Communications) mission.

It is thought that new Soviet research into mines is now developing toward missile vehicles carrying clusters of small mines and a new submarine launching system free from bubbles or any noise. There are some who say the Soviets are not very innovative and original in their endeavors. However, one should remember that in 1957 the Soviets launched the "Sputnik," and as a result President Kennedy launched a crash program for the U.S.A. to play "catch-up."

The Soviets have many more MCM vessels than the U.S.A. Their inventory of mines is also much larger. The Soviet Navy has mines that can be used to depths on the order of almost 1000 meters. Their mine-laying platforms are also more numerous and diversified.

Future Soviet Operational Philosophy

Based on an analysis of their writings, it appears that the Soviets contemplate the use of mine warfare on both a strategic and tactical basis. Strategically, in keeping with now-retired Admiral Gorshkov's "battle against the shore" philosophy, one can expect the mining of ports, harbors, and the use of mines in blockade operations. The mining of choke points and straits with the use of mines in a SLOC interdiction operation can be expected. The mine can also be used as an ASW weapon, and in an anti-SSBN role. Mining in this regard could be expected against SSBN bases, access routes, holding and launch areas.

Tactically, mine warfare would be used for base defense, anti-amphibious landing operations, ASW barriers, SLOC interdiction efforts, supporting the combat stability of ground forces from seaward, supporting of their own amphibious landing operations, anti-ship operations, and protection of their own SLOCs. Basically this means that mine warfare is considered when carrying out the majority of missions assigned to their navy. Perhaps the U.S. Navy needs a similar philosophy.

About the Author

Mr. Louis C. Gerken, President of American Scientific Corp., has been involved in the leading edge of anti-submarine warfare technology for the past 30 years. His company is a consulting leader in the fields of ASW, electronics warfare, logistics research and development, and is actively involved in the development of numerous electronic and electromechanical systems.

Mr. Gerken holds patents on various devices with both military and civilian applications, and he is the recipient of many commendations for projects conducted for the military. He has authored studies for industrial and military organizations.

Mr. Gerken holds advanced degrees in Electronics Engineering and Business Management. He holds the rank of Colonel in the U.S. Army Reserve in Research and Development. He is a graduate of the Army Command and General Staff College and Department of Defense Logistics Management College and Executive Development Courses. He lectures on subjects such as ASW, EW and logistics.

His experience is internationally founded. He has worked with the Royal Air Force, Spanish, Australian and Japanese Navies in the fields of ASW and EW—both as an engineer and as a tactician. He has been involved in new concepts in ASW with the U.S. Navy, in aviation and ships, including an evaluation of the Sea Control Ship for Admiral Zumwalt. Additionally, Mr. Gerken developed one of the first EW training simulators for the U.S. Air Force Strategic Air Command.

Professional affiliations include memberships in the Electronics Warfare Society (Old Crows), American Society of Naval Engineers, U.S. Naval Institute, Naval Submarine League, and Armed Forces Communications and Electronics Association (AFCEA).

Technical/historical books written by Mr. Gerken include *Mine Warfare, ASW vs. Submarine Technology Battle, Torpedoes* and *Airships.*